Machine Intelligence and Related Topics

MACHINE INTELLIGENCE
and Related Topics

an Information Scientist's Weekend Book

DONALD MICHIE
University of Edinburgh

with a Foreword by LORD ROTHSCHILD, FRS

GORDON AND BREACH SCIENCE PUBLISHERS
New York   London   Paris

Gordon and Breach, Science Publishers, Inc.
One Park Avenue
New York, NY 10016

Gordon and Breach Science Publishers Ltd.
42 William IV Street
London, WC2N 4DE

Gordon & Breach
58, rue Lhomond
75005 Paris

Library of Congress Cataloging in Publication Data

Michie, Donald
    Machine intelligence and related topics.

    Includes index.
    1. Artificial intelligence.   I. Title
Q335.M47   001.53′5   80-83720
ISBN 0-677-05560-9   AACR2

Contents

PART B: Sunday

Foreword

The traditional task of the writer of Forewords is to inject the expectant reader with the right blend of augustness and cliché. Such a role is not suitable in the present case.

When reading Donald Michie's book my mind was revisited by images of wartime science. At first sight there is not much in common between the use of computational skills to crack codes and the use of other skills to dismantle unexploded bombs. Yet Michie and I were both in a sense doing the same thing. In huts and sheds and back rooms a scientific generation discovered that it is a gain, not a loss, to merge the pursuit of individual excellence in a common cause, and that a good technical man should be ready to turn his hand to anything. Botanists applied themselves to anti-submarine warfare, zoologists to aircraft design, astrophysicists to mine-detection, chemists to radar. The back-room generation also discovered in themselves an attachment to their country and a resolve to promote intellectual and technical advance to a point where civilisation could not again be browbeaten.

After the war Donald Michie went to Oxford as a classical scholar, where he embarked on a career as a geneticist. I was a Cambridge biologist, and became aware of him when my colleague R.A. Fisher, the architect of modern mathematical statistics, remarked to me on an unusual doctoral dissertation which he had been examining. To be precise, he said: 'Michie is a near-genius', a remarkable tribute from a man, a near-genius himself if ever there was one, who was not given to lavishing praise on others, somewhat the reverse. Subsequently, Michie turned from a successful career in biology to the then embryonic field of machine intelligence and proceeded to build a centre of research pre-eminent in Europe.

As can be surmised from some of the later chapters, not everyone welcomed this nor even considered the goals of such work to be possible or proper. Now, however, one reads of Japan's eight-year plan to create a Fifth Generation Computer System in which 'intelligence will be greatly improved to approach that of a human being'. The issue then is not whether machine intelligence is feasible, but what share in its development is to be allowed to British as well as Japanese and American talent.

I commend this book not only to the scientific readership, which is assured, but also to those in Government with influence on such issues.

ROTHSCHILD
January, 1982

Acknowledgements

The Author wishes to thank the copyright holders for permission to re-print the previously published articles listed below in the order in which they occur in the book.

Sciencemanship (1969). *Discovery* **20**, pp. 259-60.

An Introduction to Conversational Computing (1969/70). *Mathematical Spectrum* **2**, 1, pp. 7-14.

The Bletchley Machines (1973). *The Origins of Digital Computers*, (ed. B. Randell), New York: Springer-Verlag, pp. 327-8.

Turing and the Origins of the Computer (1980). *New Scientist*, 22 February.

Bayes, Turing and the Logic of Corroboration (1976). *AISB European Newsletter*, **23**, pp. 33-36.

Heuristic Search (1971). *The Computer Journal*, **14**, 1.

King and Rook Against King (1977). *Advances in Computer Chess* **1** (M. Clarke, ed.), Edinburgh: Edinburgh University Press, pp. 30-59.

Chess with Computers (1980). *Interdisciplinary Science Reviews*, **5**, 3, pp. 215-27.

Amplifying Intelligence by Machine (1966). *New Scientist*, 29 September.

The Death of Paper (1967). *University of Edinburgh Bulletin*, **3**, 8.

Clever or Intelligent? (1969). *Cybernetics* **XII**, 4.

The Intelligent Machine (1970). *Science Journal*, **6**, 10.

Teaching a Computer to 'See' (1977). *Computer Weekly*, 3 March.

Artificial Intelligence in the Micro Age (1979). *Practical Computing* (originally published as Preserving the Vital Link of Comprehension), September.

Expert Systems (1980). *The Computer Journal*, **23**, 4.

Life with Intelligent Machines (1976). Proceedings of the Seminar on the Use of Models in the Social Sciences (ed. L. Collins), London:

Tavistock Press, pp. 101–109.

Machine Intelligence in the Cycle Shed (1973). *New Scientist*, 22 February.

Pillars of the Tabernacle (1973). *University of Edinburgh Bulletin* **10**, 4.

Peer Review and the Bureaucracy (1978). *Times Higher Education Supplement*, 4 August, p. 11.

Song and Dance Story (1979). *New Scientist*, 24 May.

Scientific Advice to Governments (1981). *New Scientist*, 23 April.

Understanding the Machine (1973). *Computer Weekly*, 5 April.

Man's Future in the Knowledge Game (1975). *Computer Weekly*, 4 September.

The Only Way Out of the Economic Pit? (1976). *Computer Weekly*, 23 September.

Pressing the Fun Button (1977). *Computer Weekly*, 10 November.

Beware SCUM – the Society for Cutting Up Machine-makers (1978). *Computer Weekly*, 29 June.

The Human Interface (1978). *Spectrum*, 2 August.

The Social Aspects of Artificial Intelligence (1980). *Micro-electronics and Society*, (ed. Trevor Jones), Milton Keynes: The Open University Press, pp. 115–43.

Preface

This is a weekend book written by an information scientist. It is also a book for information scientists to read at weekends. Both ways of taking the subtitle are intended.

It may, however, be argued that scientists do not have weekends. It is not that they work all the time. It is rather that they are incapable of scheduling those times during which they *are* working so as to fit their Creator's scheme. The important thing, however, is whether at the given moment the scientist *thinks* it is the weekend. Thus, if he is at the cinema you can be sure that he imagines it to be Saturday, if on a family picnic then Sunday. For myself, the arrival of the weekend, or at least of the idea of the weekend, is reliably signalled by an outburst of various self-indulgences. One of these is the composition of bits and pieces about science. This has nothing whatsoever to do with the activity known as 'scientific publication' — an intrinsic if burdensome and much postponed part of the working life, devoted to putting new results on record.

Among the writings assembled here, one or two are records of this last kind, but they nonetheless belong to the weekend for some other reason. The rather technical account of a chess end-game on an infinite board has been included because the style is popular and the motivation was not strictly of the working week. The circumstance is explained in the note preceding the article, and most of the articles have been supplied with prefatory notes of one sort or another.

The bits and pieces have been apportioned between Saturday and Sunday on the following criterion. At St. Peter's gate my turn will come, as to all other souls, to answer those feared accountant's questions concerning the time mis-spent. 'And why did you write this?' St. Peter will ask, indicating yet another wretched offering from a life-time's weekend scribbling. If I can reply 'Part, holy Sir, of my life as a scientist' while looking steadily in his eye, then it was surely a Saturday composition. Otherwise I have put it in the last part of this book.

DONALD MICHIE

xi

Preamble

Introductory Note to *Sciencemanship*

Superficially a Sunday piece, this article claims re-classification under St. Peter's principle explained in the Preface. For if it is not part of a scientist's duty to pass on tips to other strivers, then he has no duty. I have, however, skirted the issue by placing it apart as common preamble to all that follows.

As an elder, I look back 30 years towards Michie the Younger, author of *Sciencemanship*. I see a committed man impatient of his seniors and of their cosy world. It gives me a start to realise that, relativistically speaking, I am now of their number, although he is no longer here to judge me. How *would* he judge me? The question seems pertinent, and will no doubt be answered in my memoirs.

Sciencemanship

The latest book on How to be a Good Scientist is by Professor D.J. Ingle of Chicago, USA, and has the title, *Principles of Biological and Medical Research*. At the end I found myself wondering why it had so little to add to the admirable earlier studies of Beveridge (*The Art of Scientific Investigation*) and of Wilson (*An Introduction to Scientific Research*). Perhaps it is not so easy to break fresh ground in this field. However, I am not entirely convinced of this.

Murphy's Law

I think that the deepest and most durable impression which the research man's mind sooner or later receives is how unexpectedly, how unjustly, how distressingly difficult it seems to be to discover or prove anything at all. The research worker would be spared much early perplexity if his formal instruction included a sound treatise on Murphy's Law.

This important Law is described by H.B. Brous Jr. in the September number of *Astounding Science Fiction* as stating, 'If anything can go wrong, it will.' Research men among my readers will instantly recognise the truth and generality of this Law, even if they have not previously come across its verbal formulation. But having recognised it, what to do about it? Ingle's book, in common with those of Beveridge and of Wilson, has no definite proposals to make.

Here, then, is a suggestion, offered as a stimulus to others interested in this uncharted territory. The moment to take account of Murphy's Law is clearly when you are planning a new investigation. You have worked out how much material will theoretically give you the required amount of information. We will call this theoretical estimate x. x may be the number of rats to be treated, or of acres to be sown, or of soil samples to be collected, and so on. You then attempt to make rational allowance for all the things which might go wrong. While judging any specified mishap to be highly improbable, you might yet consider that the joint effect of all the

4

improbable mishaps might amount to say, a possible 30% wastage. You therefore decide to budget for 1.43 times the theoretical estimate (after 30% wastage, 1.43x becomes x), and the multiplier you use (in this case 1.43) I call the Rational Multiplier, R.

$M = R^2$

It is at this stage that we usually finalise our plans, and live to regret it. It turns out that although some of the possible hazards did not materialise, we had forgotten that a proportion of the rats might have fatal convulsions on hearing a whistling kettle, and that a colleague might mistake some of the clearly labelled organs stored in the refrigerator for goldfish food, and act accordingly. It is before any of this happens that Murphy should be consulted. Having quizzically surveyed the wreckage of many an experiment, I assert that the needed prophylactic lies in the use of the *Murpheian* multiplier, M, in place of R, to which it is related by the simple expression $M = R^2$. In our hypothetical case, supposing that the inexperienced (entirely theoretical) man would procure 100 rats from the dealer or animal house, the 'rational' man would procure 143, but Murphy would procure 204.

The expression $M = R^2$ rests on more than empiricism. It was derived, with the aid of my colleague, Anne McLaren, from certain theoretical considerations. These involve the idea that the Rational Multiplier depends on the number of discrete risk-bearing operations into which the total experiment can be broken down. If the rationally foreseeable risk attached to each of these is assumed to be accompanied by an independent, unforeseen, Murpheian risk of equal magnitude, then the above equation follows. Of course, the assumption is crudely approximate. But it is a beginning.

Necessity of Idleness

Beveridge has emphasised the need of the research man to lie fallow for periods of time, and quotes J. Pierpont Morgan as saying, 'I can do a year's work in nine months, but not in twelve months.' Unfortunately, he offers no concrete suggestions. A former colleague at one time installed a camp bed in his lab so that he could lie down when he felt tired or lazy. His Departmental Head disapproved, but I think that the idea is interesting.

An allied problem is the Visitor Menace. I knew a famous man of science who, when a self-invited visitor was in the offing, would retire to the cloakroom. He took with him his papers and books, and emerged only when the 'all clear' was sounded. I find no reference to the cloakroom manoeuvre in Ingle's book, and Beveridge and Wilson are also without practical recommendations.

The visitor menace is an expression of a general, and truly paralysing, affliction which overtakes most research men in their mature years. This is the Earnestness of being Important. Ingle says, 'The early years in the laboratory are the golden years for many scientists. After he becomes known, the volume of mail, telephone calls, number of visitors, organisational activities, including committees by the dozens, and demands for lectures, reviews, and community activites grow insidiously and will destroy the creativity of the scientist if unopposed.' But how to oppose them? In a delightful essay on 'Heads of Research Laboratories' (translated in SCR *Soviet Science Bulletin*, 1957, vol. 4, p 1–6) the Soviet Academician A.L. Kursanov utters a similar warning: 'They come to us, these administrative commitments, of their own accord in the fullness of time, and the less we want them the sooner they come.' But he also fails to advance a concrete plan of self-defence for the research man.

Five Principles

If only to start the ball rolling, here are five principles of evasion, not yet tried and tested, but perhaps deserving of trial.

   1) No committees.
   2) No refereeing.
   3) No editing.
   4) No book-reviewing.
   5) No invited papers.

Special dispensation can possibly be granted for anything for which the hard-up research man can get sufficiently well paid (for example, reviewing Ingle for Discovery). The fifth item is the least obvious, but rather interesting. I added it recently when I had been going through my collection of reprints of the scientific papers of others, to discard those which I felt I could do without. At the end, I found to my surprise that my reject pile contained a high proportion of papers which had been delivered by invitation to some conference or symposium. I looked at them again. Many were by highly gifted authors on subjects of great interest to me. I still did not want them. The proportion of invited papers in my non-reject pile was small.

The clue probably lies in the recipe which one tends to follow for putting together an invited paper for a special occasion. The recipe is hash and waffle. By 'hash' I really mean *re*-hash of results which have in the main already been published somewhere else. The concoction can be diverting and informative for one's listeners. But it seems that hash and waffle is a dish that does not keep.

This is not to suggest that published symposia and colloquia are not of immense value for the advance of science. Quite the contrary: the explosive expansion today of almost every sector of the scientific front makes a vital necessity of any and every means of keeping scientific workers in touch with each other, and with the latest advances in their own and neighbouring fields. The scientist who helps to perform this service richly deserves the gratitude and admiration of his fellows. But let him not think that he thereby necessarily makes an original and lasting contribution to knowledge. If *this* happens to be his ambition, there can be no compromise. He must be prepared ruthlessly to disembarrass his thoughts and time-tables from every preoccupation other than his central quest.

Chap Rotation

I once worked in an applied research outfit which, among other peculiar practices, operated a sort of rotation of crops, or, rather, rotation of chaps. Once in every while — I do not now recall whether it was once in six or seven or eight weeks — each man was banished to a small room for a week, in which his only duty was to sit and muse. No one asked at the end of the week, 'Did you have any bright ideas?' for this in itself might damp the muse. He was only asked to abstain from all routine work during that week. In exchange, the exile had arbitrary powers to commandeer any of the outfit's equipment or labour force if he wished to test his latest bright idea.

Some heads of research teams may look askance at this scheme. To those who are tempted to try it in their lab, I should emphasise the following. It must be made very clear that the exile who spends an apparently barren week with his feet on the table reading the comics gains the same merit in the eyes of the team and its leader as the one who emerges to suggest six new experiments and a modification of the Second Law of Thermodynamics. Otherwise the whole point is lost.

Browsing

The rotation of chaps is only one of many possible devices for recharging the research man's mental batteries. The necessity of recharging is eloquently stated by Kursanov, employing a different metaphor: 'A scientist is not a balloon, to reach a certain height and remain there for a long while on account of the material it was once filled with. He is "heavier than air", more like an aeroplane, which has to keep going to maintain height or climb.' It is well known that height on average is not maintained. Beveridge cites Lehman's figures for output at different ages. Taking the decade of life 30–39 as 100, the output for the years 20–29 was 30%–40%; for 40–49, 75%; for 50–59, about 30%. Assuming that the slow start is due to lack of knowledge and experience, is the later decline entirely due to biological ageing? I think not. Two features of a young scientist's life at once occur to me which tend to disappear with time and which may be important. One is browsing and the other is fairly frequent change of work and surroundings.

What senior scientist can be found sitting all day in the library looking through research periodicals because he has nothing else particular to do? And what research student does not from time to time do just this? As for change of work, Beveridge mentions the case of Ostwald, who successfully rejuvenated his mind by this means when he was over fifty years of age. In this connection a proposal made by Kursanov's countryman, the nuclear physicist Peter Kapitsa, deserves attention. Kapitsa intends his suggestion for adoption in Russia, but there seems no obvious reason why it should not be applied more widely.

Combat Forces

His idea is the setting up of *ad hoc* 'mobile combat forces', each to be regarded 'not as a permanent institution but as one set up to tackle a given problem over a period of months or years'. Such a force would consist of scientists drawn from a number of different specialities, each with some special angle on the problem to be solved. After the successful solution of the problem, the combat force would be dissolved and its members would return to the permanent departments or institutes from which they had been recruited, or some of them might join new combat forces.

In fact, something like this occurred in Britain during the war, but with a measure of compulsion inadmissible in peacetime. Apart from the gain in efficiency, I see a valuable psychological advantage in such a scheme. It would enable even the senior research man to reverse his trend towards stagnation, for the scientific mind is more like a medicine than a wine: it should be well shaken before use.

Many readers will have other, and better, suggestions than those which have been aired here. But enough, I think, has been said to show how many and how inviting are the paths in which Ingle has failed to tread.

PART A

Saturday

Introductory Note to Chapter 1

Conversational Computing, written in 1969, was the script of a demonstration given to a meeting of the British Association for the Advancement of Science. Remote conversational use of computers was then so new as to be regarded as miraculous. The teletype at Dundee was connected to our Edinburgh machine via the public telephone network — a novel indecency requiring special dispensation from the GPO — with stand-by provision for switching if need be to a computer at Stirling. This was just as well. A note handed to the platform after my first few words informed me that the machine at Edinburgh had gone out of commission. By the time I turned to the teletype, so had the Stirling computer. Back-room colleagues at the two ends of the Dundee–Edinburgh link saved the day and the demonstration proceeded without apparent hitch. Both lecturer and audience were by then under a common misapprehension as to where exactly the conversation was being conducted, which was by now back in Edinburgh.

CHAPTER 1

An Introduction to Conversational Computing

This article is based on a particular style of computing, the so-called 'conversational mode', in which the user is free to conduct a rapid-fire dialogue with the machine, choosing his next response from moment to moment in the light of the latest output of the machine and of his own changing view of the situation.

In order to bring to life the essential nature of interactive computing, and the resources of a modern conversational computing language, I shall introduce the notion of a computing facility as a workshop for building internal (abstract) models of external (concrete) reality. Notice that I have framed my definition so widely that it could actually cover much of the activity which goes on inside our own skull-boned computing machines and which we call 'thought'. I have done this deliberately, since the restriction of digital computing to the processes of arithmetic is scarcely more than a historical accident, as artificial as if we were to confine the terms of reference of a metal workshop to operating upon nuts and bolts. This is not to say that nuts and bolts are not important.

I spoke above of a 'computing facility' rather than a 'computer'. This is because a naked computer, in the form in which it leaves the factory, can hardly be said to be a machine at all. It is only a potential machine. Which particular machine it will become out of an infinite range of possibilities is determined by the particular program with which it is loaded. I am going to assume that our computer has been loaded with the compiler program for a modern conversational computing language. A compiler is a program which translates from a computing language, which is oriented towards the human user's natural style of expressing himself, into the machine's own internal language. I shall further assume that this language is a truly mathematical (not just a numerical) one. In other words, in the general command schema

$$f(x) \rightarrow y$$

('apply the function f to the object x and call the result y') the user is free to define an f for x of any type whatsoever — integer, real, Boolean, word, string, list, array, record, set, group, ring, graph, etc. — and the type of y can be similarly unrestricted. Demanding this degree of mathematical generality narrows the range of programming languages from which to choose. For the illustrations which follow I shall use Burstall and Popplestone's language POP-2 with which I am most familiar.

In the real world we expect a workshop to contain

— materials
— structures
— machines (including hand-tools).

We use a machine to operate on one or more structures and/or materials to make one or more new structures. The new structure may or may not itself be a machine. Thus we apply a riveter to a pot, a handle, and some rivets to make a saucepan. A saucepan is itself a machine which, when applied to heat, water, and the contents of an appropriate packet, will make soup. The abstract world of computing correspondingly contains

— store
— data structures (e.g., lists, arrays, etc.)
— functions.

Thus we might apply a list-pairing function to a list of French words and a list of English words to make an association list or dictionary. To get the feel of the language the following POP-2 function for doing this is worth looking at in detail:

```
FUNCTION PAIROFF LIST1 LIST 2;
    VARS NEWLIST;
    IF NOT (LENGTH(LIST1) = LENGTH(LIST2)) THEN
        PR('UNEQUAL') EXIT;
    NIL → NEWLIST;
NEXT: IF LIST1.NULL THEN NEWLIST EXIT;
    (LIST1.HD: :[%LIST2.HD%]): :NEWLIST → NEWLIST;
```

LIST1.TL → LIST1; LIST2.TL → LIST2;
GOTO NEXT
END;

The first line is synonymous with 'LET PAIROFF(LIST1,LIST2) BE . . .',
i.e., it names the function and introduces formal names for its arguments.
In the next line we declare a local variable 'NEWLIST', which will be used
to contain the result of the function. Next comes a test for an error condition (list lengths unequal) which, if positive, causes exit from the function
with a printed message 'UNEQUAL'. Now the real business starts, with the
creation of an empty association list, which is assigned to the variable
NEWLIST. The infixed operator ':: ' is a joining function, and HD and
TL are functions which select the head and the tail of a list respectively.
Thus [FEE FI FO].HD (a shorthand way of writing HD([FEE FI FO])
is FEE.[FEE FI FO].TL is [FI FO]. [FEE FI FO].HD : : [FI FO FUM]
is [FEE FI FO FUM]. In the 'pairoff' example above, the brackets of one
of the lists have been decorated with per cent signs. This is done when the
elements of the list are expressions, which need to be evaluated, as opposed
to constants such as FEE or 2. [% 'CHARGE', 'IS', SQRT (X), 'POUNDS'
%] is therefore equivalent, if x happens to have the value 4 at the time, to
[CHARGE IS 2 POUNDS]. We now repeatedly go round the loop, taking
the heads off LIST1 and LIST2, joining these heads in pairs and chaining
each pair in turn to the top of NEWLIST. When LIST1 and LIST2 are ex-
hausted exit occurs with NEWLIST as the value of the function.

Suppose I type this function definition on a conversational terminal
and follow it with

VARS FRENCH ENGLISH VOCAB;
[VACHE CHIEN CHAT HOMME MANGER DORMIR] → FRENCH;
[COW DOG CAT MAN EAT SLEEP] → ENGLISH; (square brackets
are used by convention to enclose lists).

I can now construct a French–English vocabulary by

PAIROFF(FRENCH,ENGLISH) → VOCAB; and take a look at it
by typing
VOCAB ⇒
(the symbol ⇒ is a print command).

The teletype will reply

**[[DORMIR SLEEP] [MANGER EAT] [HOMME MAN]
[CHAT CAT] [CHIEN DOG] [VACHE COW]],

Note that in POP-2 the machine always prefaces its response to the ⇒ print command with the symbols **. If I don't like the reversal of the original order of items I can modify my 'pairoff' definition, for example by giving REV(NEWLIST), rather than NEWLIST, as the result. This illustrative exercise may appear at first sight to be a little vacuous, since all that PAIROFF does is to replace 'row pairing' by 'column pairing'. The usefulness of simple manipulations of this kind is seen if we now imagine writing a look-up function for French–English translation, such as the following:

```
        FUNCTION LOOKUP ITEM ASSOCLIST;
  LOOP: IF ASSOCLIST.NULL THEN 'UNKNOWN' EXIT;
      IF ASSOCLIST.HD = ITEM THEN
          ASSOCLIST.TL.HD EXIT;
      ASSOCLIST.TL → ASSOCLIST;
      GOTO LOOP
  END;
```

We would test this in the above example by typing, say,

LOOK UP(HOMME,VOCAB) ⇒, receiving the reply
**MAN, or
LOOKUP(FEMME,VOCAB) ⇒, with the reply
**UNKNOWN,

If we now update our list by

[FEMME WOMAN] : : VOCAB → VOCAB;

we can again type

LOOKUP(FEMME,VOCAB) ⇒, this time with a different result:
**WOMAN,

Now let us press the workshop analogy a bit harder: let me imagine myself in a workshop for mixing and spraying paints. My task is to provide myself with a few dozen pots, to fill the first few with some starting colours, say red, blue, and yellow, and to use a paint mixer and a paint-spray to apply paints of desired colours to specified objects — say pegs. First I set up a couple of dozen empty pots, by creating a one-dimensional array:

        VARS POT;
        NEWARRAY ([%1,24%] , INITIAL) → POT;

'NEWARRAY' is a constructor function of two arguments, the first — '[%1,24%]' — a list of lower and upper bounds and the second — 'INITIAL' — a function which assigns starting values to the cells of an array. If we wanted the cells of a paintpot array to start empty we might previously define 'INITIAL' as follows:

        FUNCTION INITIAL N; 'BLANK' END;

to produce the result 'BLANK' for all values of N.
    To check that all is well we now type

        POT(2) ⇒

and on receiving the answer

        **BLANK,

we proceed to fill the first three pots.

        'RED' → POT(1);
        'BLUE' → POT(2);
        'YELLOW' → POT(3);

Typing the statement

        POT(2) ⇒ now produces the result
        **BLUE.

The next need is for a colour-mixing machine, which samples paint from two pots and puts a blend of the two into a third pot. For this purpose I have invented some simple laws of colour blending and have designed the mixing machine accordingly:

```
FUNCTION PAINTMIX I J K;
    IF NOT (POT(K) = 'BLANK') THEN 'FULL' ⇒ EXIT;
    VARS COLOUR1 COLOUR2;
    POT (I) → COLOUR1; POT(J) → COLOUR2;
    IF COLOUR1 = 'MESS' OR COLOUR2 = 'MESS' THEN 'MESS'
        → POT(K) EXIT;
    IF COLOUR1 = COLOUR2 OR COLOUR1 = 'BLANK' THEN
        COLOUR2 → POT(K) EXIT;
    IF COLOUR2 = 'BLANK' THEN COLOUR1 → POT(K) EXIT;
    IF COLOUR1 = 'BROWN' OR COLOUR2 = 'BROWN' THEN
        'BROWN' → POT(K) EXIT;
    IF COLOUR1 = 'BLUE' AND COLOUR2 = 'YELLOW' THEN
        'GREEN' → POT(K) EXIT;
    IF COLOUR1 = 'RED' AND COLOUR2 = 'BLUE' THEN
        'PURPLE' → POT(K) EXIT;
        . . .
        . . .
        . . .
    ELSE 'MESS' → POT(K) CLOSE
END
```

The details need hardly concern us, but the intended effect is to make plausible colour-products from any mixture of which one component is either brown or a primary colour, and otherwise to classify the result as a 'mess'.

Now to check that the mixer is working:

```
PAINTMIX(1,2,4);
POT(4) ⇒
**PURPLE,
PAINTMIX(3,4,4);
**FULL,
PAINTMIX(2,3,5); PAINTMIX(1,3,6);
```

POT(5) ⇒
**GREEN,
POT(6) ⇒
**ORANGE,

Remember that the machine's responses are determined by the information written into the definition of the PAINTMIX function. The user is in a position similar to someone carrying out a physical experiment on a tool of his own design, except that he has complete knowledge, in principle, of the laws governing the results, and the freedom to change these laws at will.

Now to construct a paintspray, or, to revert to the language of abstractions, to define a 'paintspray' function. This is an abstract model of a machine which when applied to a paint and a peg transfers the former to the latter's head

FUNCTION PAINTSPRAY PEG PAINT;
    PAINT → PEG.HD
END

Create some objects for painting: VARS PEG1 PEG2 PEG3 PEG3; and give them 'blank', i.e., uncoloured, heads to start with.

[BLANK PEG] → PEG1;
[BLANK PEG] → PEG2;
[BLANK PEG] → PEG3;
[BLANK PEG] → PEG4;

Now we try painting them:

PAINTSPRAY(PEG1,POT(4));
PEG1 ⇒
**[PURPLE PEG],
PAINTMIX(1,4,8);
PAINTSPRAY(PEG2,POT(8));
PEG2 ⇒
**[MAGENTA PEG]
PAINTMIX(5,6,7);

```
POT(7) ⇒
**MESS, so we empty pot 7; 'BLANK' → POT(7); and re-fill it;
PAINTMIX(3,4,7);
POT(7) ⇒
**BROWN
```

We get tired of spelling out 'PAINTMIX' each time and give it a shorter name:

```
VARS MIX; PAINTMIX → MIX; now apply the new name:
    MIX(3,7,9);
```

We now need an instrument for displaying the contents of the first *N* pots:

```
FUNCTION PRINTPOTS N;
    VARS COUNT; 0 → COUNT
NEXT: IF COUNT = N THEN NL(3) EXIT;
    COUNT + 1 → COUNT;
    NL(2); PR(COUNT); SP(1); PR(POT(COUNT));
    GOTO NEXT
END;
```

'SP(N) means 'print *N* blank spaces' and 'NL(N)' means 'do carriage return and line feed *N* times'.

We try this function out. The command provokes the following response:

```
PRINTPOTS(10);
    1 RED
    2 BLUE
    3 YELLOW
    4 PURPLE
    5 GREEN
    6 ORANGE
    7 BROWN
    8 MAGENTA
    9 BROWN
   10 BLANK
```

and again:

        MIX(1,8,10); MIX(2,8,11); MIX(3,8,12);
        MIX(2,5,13); MIX(3,13,14); MIX(1,6,15);
        PRINTPOTS(16);
            1 RED
            2 BLUE
            3 YELLOW
            4 PURPLE
            5 GREEN
            6 ORANGE
            7 BROWN
            8 MAGENTA
            9 BROWN
            10 RED
            11 PURPLE
            12 BROWN
            13 TURQUOISE
            14 GREEN
            15 VERMILION
            16 BLANK

Clearly any amount of elaboration and further fun can be extracted from this nursery exercise, without adding much didactic value. But before leaving it, there is just one more example worth exhibiting of the parallelism between abstract model-building and building real models in real workshops. Suppose I have a tool or machine which normally requires two inputs to work on, just as a paintspray needs to have its back end applied to a paint source and its front end to a target. What happens if I apply it to just one of its inputs, and in some way freeze this input into its structure (as I might join a paintspray to a paintpot and then weld the join solid)? The commonsense answer is that the result will be a machine of correspondingly restricted application, no longer a paintspray but, for example, a redspray, which now requires only one input (its target) to be supplied for it to do its work. There is not only commonsense but mathematical sense in this, for there is a real basis for regarding, say *add* (3,1) → 4 as proceeding via the creation of an intermediate function '*add* 1' (i.e., the successor function in this example) which is then applied to 3 to give the final answer. The first of these two successive steps is called in POP-2 'partial application' of a function.

To illustrate using POP-2, we could say

    ADD(%1%) → SUCCESSOR; (decoration of the argument brackets
                          with 'per cent' signs serves as a signal
                          that partial application is being used)

    SUCCESSOR(3) ⇒
    **4,

or in terms of the paintshop

    VARS SPRAY1 SPRAY 2;
    PAINTSPRAY(%POT(1)%) → SPRAY1;
    PAINTSPRAY(%POT(6)%) → SPRAY2;

Now I can do the following

    SPRAY1(PEG4); SPRAY2(PEG3);
    PEG4 ⇒
    **[RED PEG],
    PEG3 ⇒
    **[ORANGE PEG];

Notice that if I now change the contents of pot 1 to green, e.g.,

    'GREEN' ⇒ POT(1);

the SPRAY1 function will continue to produce 'RED' and will not switch to 'GREEN'. The device of partial application will allow us to freeze into the function the value which the relevant part of its environment (in this case pot 1) has *at the time* – a most useful property when manipulating abstract models of a changing world.

The use of computers for model-building in industrially important applications is being extended today to the simulation of entire chemical plants, traffic systems, telephone networks, aeroplane structures and so forth, requiring computer programs costing dozens of man-years to write, test and maintain in use. At a simpler level a very wide range of non-numerical computer uses can be regarded as paralleling familiar categories in the outside world, in a sense which I hope my toy example has helped to make clear.

Introductory Note to Chapters 2, 3 and 4

I placed the next three articles together after a visit paid to Professor Max Newman in his retirement. His role in the development of the COLOSSUS proto-computers figures in the first of them. Among reminiscences we spoke of Alan Turing. Newman commented that possibly Turing's most influential contribution at Bletchley was that he recast the methodology of Bayesian probabilistic inference into improved and more easily computable form. The Official Secrets Act does not permit excursion into the cryptanalytic uses to which Turing's innovations were applied, but the third article of the batch popularizes the nature of this particular idea.

CHAPTER 2
The Bletchley Machines

During the war the Department of Communications of the British Foreign Office was housed at Bletchley Park, Buckinghamshire, where highly secret work on cryptanalysis was carried out. As part of this work various special machines were designed and commissioned, the early ones being mainly electromechanical, the later ones electronic and much closer to being classifiable as program-controlled computers. Even now relatively few details of these machines have appeared in the published literature (on which this account has itself been based) and a really adequate description and assessment of these machines will have to await a relaxation of the security classification which is still imposed on them.

The first of the electromechanical machines, the 'Heath Robinson', was designed by Wynn-Williams at the Telecommunications Research Establishment at Malvern. At Bletchley one of the people with influence on design was Alan Turing. The machine incorporated two synchronised photoelectric paper tape readers, capable of reading 3,000 characters/sec. Two loops of 5-hole tape, typically more than 1,000 characters in length, would be mounted on these readers. One tape would be classed as 'data', and would be stepped systematically relative to the other tape, which carried some fixed pattern, by differing in length from it by, for example, one character. Counts were made of any desired Boolean function of the two inputs. Fast counting was performed electronically, and slow operations, such as control of peripheral equipment, by relays. The machine, and all its successors, were entirely automatic in operation, once started, and incorporated an on-line output teleprinter or typewriter.

Afterwards, various improved 'Robinson's' were installed, including the 'Peter Robinson', the 'Robinson and Cleaver' and the 'Super Robinson'. This last one was designed by T.H. Flowers in 1944, and involved four tapes being driven in parallel. Flowers, like many of the other engineers involved in the work, was a telephone engineer from the Post Office Research Station.

26

The electronic machines, known as the COLOSSI, were developed by a team led by Professor M.H.A. Newman, who started the computer project at Manchester University after the war. Other people directly involved included T.H. Flowers, A.W.M. Coombs, S.W. Broadbent, W. Chandler, I.J. Good, and D. Michie. During the later stages of the project several members of the US armed services were seconded at various times to work with the project for periods of a year or more.

Flowers was in charge of the hardware side, and in later years designed an electronic telephone exchange. On his promotion, his place was taken by Coombs, who in post-war years designed the time-shared trans-Atlantic multi-channel voice communication cable system. After the war, I.J. Good was for a time associated with the Manchester University computer project, and Coombs and Chandler were involved in the initial stages of the design of the ACE computer at the National Physical Laboratory, before building the MOSAIC computer at the Post Office Research Station. Alan Turing was apparently not directly associated with the design of the COLOSSUS machine, but with others provided the requirements that the machines were to satisfy. It has, however, since been claimed by Good that Newman, in supervising the design of the COLOSSI, was inspired by his knowledge of Turing's 1936 paper.

In the COLOSSUS series almost all switching functions were performed by hard valves, which totalled about 2,000. There was only one tape, the data tape. Any pre-set patterns which were to be stepped through these data were generated internally from stored component patterns. These components were stored in ring registers made of thyrotrons and could be set manually by plug-in pins. The data tape was driven at 5,000 characters/sec. In the Mark 2 version of the machine an effective speed of 25,000 characters/sec. was obtained by a combination of parallel operations and short-term memory. Boolean functions of all five channels of pairs of successive characters could be set up by plug-board, and counts accumulated in five bi-quinary counters.

The first COLOSSUS was installed by December 1943, and was so successful that three Mark 2 COLOSSI were ordered. By great exertions the first of these was installed before D-day (6th June, 1944). By the end of the war about ten COLOSSI had been installed, and several more were on order.

CHAPTER 3
Turing and the Origins of the Computer

Everyone who knew him agreed that Alan Turing had a very strange turn of mind. To cycle to work at Bletchley Park in a gas mask as protection against pollen, or to chain a tin mug to the coffee-room radiator to insure against theft, struck those around him as odd. Yet the longer one knew him the less odd he seemed after all. This was because all the quirks and eccentricities were united by a single cause, the last that one had expected, namely a simplicity of character so marked as to be by turns embarrassing and delightful, a schoolboy's simplicity, but extreme and more intensely expressed.

When a solution is obvious, most of us flinch away. On reflection we perceive some secondary complication, often a social drawback of some kind, and we work out something more elaborate, less effective, but acceptable. Turing's explanation of his gas mask, of the mug chaining, or of other startling short cuts was 'Why not?', in genuine surprise. He had a deep-running streak of self-sufficiency, which led him to tackle every problem, intellectual or practical, as if he were Robinson Crusoe. He was elected to a fellowship of King's College, Cambridge, on the basis of a dissertation on the 'Central limit theorem of probability' which he had re-discovered from scratch. It seemed wrong to belittle so heroic an achievement just on the grounds that it had already been done!

Alan Turing's great contribution was published in 1937, when he was 25. While wrestling Crusoe-like with a monumental problem of logic, he constructed an abstract mechanism which had in one particular embodiment been designed and partly built a century earlier by Charles Babbage – the analytical engine. As a purely mathematical engine with which to settle an open question – the decidability problem – Turing created a formalism which expressed all the essential properties of what we now call the digital computer. This abstract mechanism is the Turing machine. Whether or not any given mathematical function can, in principle, be evaluated was shown by Turing to be reducible to the question of whether

28

a Turing machine, set going with data and an appropriate program of computation on its tape, will ever halt. For a long time I thought that he did not know about Babbage's earlier engineering endeavour. In all the talk at Bletchley about computing and its mathematical models, I never heard the topic of Babbage raised. At that time I was quite ignorant of the subject myself. But according to Professor Brian Randell's paper 'The Colossus', delivered to the 1976 Los Alamos Conference on the History of Computing, T.H. Flowers 'recalls lunch-time conversations with Newman and Turing about Babbage and his work'. However that may be, the isolation and formal expression of the precise respect in which Babbage's machine could be described as 'universal' was Turing's.

The universal Turing machine is the startling, even bizarre, centre-piece of the 1937 paper 'On computable numbers with an application to the Entscheidungs-problem'. Despite its title, the paper is not about numbers in the restricted sense, but about whether and how it is possible to compute *functions*. A function is just a (possibly infinite) list of questions paired with their answers. Questions and answers can, of course, both be encoded numerically if we please, but this is part of the formalities rather than of the essential meaning.

For any function which we wish to compute, imagine a special machine to be invented as in Figure 3.1. It consists of a read-write head, and a facility for moving from one field ('square' in Turing's original terminology) of an unbounded tape to the next. Each time that it does this it reads the symbol contained in the corresponding field of the tape, either a '1' or a '0' or a blank. This simple automaton carries with it, in its back pocket as it were, a table of numbered instructions ('states' in Turing's terminology). A typical instruction, say number 23 in the table, might be: 'if you see a 1 then write 0 and move left; next instruction will be number 30; otherwise write a blank and move right; next instruction will be number 18'.

To compute $f(x)$, – square-root (49), say, – enter the value of x in binary notations as a string of 1s and 0s on the tape, in this case '110001', which is 49 in binary. We need to put a table of instructions into the machine's back pocket such that once it is set going the machine will halt only when the string of digits on the tape has been replaced by a new one corresponding precisely to the value of $f(x)$. So if the tape starts with 110001, and the table of instructions has been correctly prepared by someone who wishes to compute square roots to the nearest whole number, then when the machine has finished picking its way backwards and for-wards it will leave on the tape the marks '111', the binary code for 7.

General Computations

When f = square-root, we can well imagine that a table of instructions can be prepared to do the job. But here is an interesting question: how do we know this? Could this be knowable in general? Could a systematic proce-dure be specified to discover for every given function whether it is or is not Turing-computable, in the sense that a table of instructions could or could not be prepared?

In the process of showing that the answer is 'No', Turing generalised the foregoing scheme. He imagined an automaton of the same kind as that already described, except that it is a general purpose machine. If we want it to compute the square-root we do not have to change its instruction table. Instead we merely add to the tape, alongside the encoding of the number whose square-root we want, a description of the square-root machine — essentially just its table of instructions. Now what is to stop the general-purpose machine obeying the symbols of this encoding of the square-root machine's instruction table? Plenty! the astute reader at once replies. This new automaton, as so far described, consists again just of a read-write head. It has no 'brain', or even elementary understanding of what it reads from the tape. To enable it to *interpret* the symbols which it encounters, another table of instructions must again be put into its back pocket — this time a master-table the effect of which is to specify a lan-guage in the form of rules of interpretation. When it encounters a descrip-tion, in that language, of any special-purpose Turing machine whatsoever, it is able, by interpreting that description, faithfully to simulate the opera-tions of the given special-purpose machine. Such a general-purpose auto-maton is a universal Turing machine. With a *language* in its back pocket, the machine is able to read the instructions 'how to compute square roots', then the number, and after that to compute the square root.

Using this construction, Alan Turing was able to prove a number of far-reaching results. There is no space here to pursue these. Suffice it to say that when mathematicians today wish to decide fundamental questions concerned with the effectiveness or equivalence of procedures for func-tion-evaluation, or with the existence of effective procedures for given functions, they still have recourse to the simple-minded but powerful formal construction sketched above.

In practical terms the insights derivable from the universal Turing machine (UTM) are as follows. The value of x inscribed on the tape at the start corresponds to the data-tape of the modern computing set-up. Almost as obviously, the machine-description added alongside corresponds to a *program* for applying f to this particular x to obtain the answer. What then is the table of instructions which confers on the UTM the ability to interpret the program? If the computer is a 'naked machine' supplied by a manufacturer who provides only what is minimally necessary to make it run, then the table of instructions corresponds to the 'order-code' of that machine. Accordingly the 'machine description' appropriate to square-root is a program written in the given order-code specifying a valid procedure for extracting the square-root. If, however, we ask the same question after we have already loaded a compiler-program for, let us say, the programming language Algol-60, then we have in effect a *new* universal Turing machine, the 'Algol-60 machine'. In order to be interpretable when the machine runs under this new table of instructions, the square-root program must now be written, not in machine code, but in the Algol-60 language. We can see, incidentally, that indefinitely many languages, and hence different UTMs, are constructible.

There are various loose ends and quibbles. To head off misunderstanding I should add that the trivial example 'square-root' has been selected only for ease of exposition: the arguments hold for arbitrarily complicated problems. Secondly, what has been stated only applies, strictly, to computers with unbounded memory. Thirdly, the first thing that a modern machine ordinarily does is to 'read in' both data and program, putting the contents of the Turing 'tape' into memory. The Turing machine formalism does not bother with this step since it is logically immaterial whether the linear store ('tape') is to be conceived as being inside or outside: because it is notionally unbounded, it was doubtless easier originally to picture it as 'outside'!

From the standpoint of a mathematician this sketch completes the story of Turing's main contribution. From the point of view of the information engineer such as myself, it was only the beginning. In February 1947 Alan Turing delivered a public lecture at the London Mathematical Society. In it he uttered the following words:

It has been said that computing machines can only carry out the purposes that they are instructed to do . . . But is it necessary that they should always be used in such a manner? Let us suppose that we have set up a machine with certain initial instruction tables, so constructed that these tables might on occasion, if good reason arose, modify these tables. One can imagine that after the machine had been operating for some time, the instructions would have been altered out of recognition, but nevertheless still be such that one would have to admit that the machine was still doing very worthwhile calculations. Possibly it might still be getting results of the type desired when the machine was first set up, but in a much more efficient manner. In such a case one could have to admit that the progress of the machine had not been foreseen when its original instructions were put in. It would be like a pupil who had learnt much from his master, but had added much more by his own work. When this happens I feel that one is obliged to regard the machine as showing intelligence. As soon as one can provide a reasonably large memory capacity it should be possible to begin to experiment on these lines.

Ten years were to pass before the first experiments in machine learning by Arthur Samuel at IBM, and 35 years before conceptual and programming tools have made possible the experimental assault which is gathering force today along the Turing line. For consider modification not only of the data symbols on the UTM tape but also of the machine-description symbols — modification *of* the program *by* the program! My own laboratory constitutes one of the resources dedicated to this 'inductive learning' approach.

In a particular sense, Alan Turing was anti-intellectual. The intellectual life binds its practitioners collectively to an intensely developed skill, just as does the life of fighter aces, of opera stars, of brain surgeons, of yachtsmen, or of master chefs. Strands of convention, strands of good taste, strands of sheer snobbery intertwine in a tapestry of myth and fable to which practitioners meeting for the first time can at once refer for common ground. Somewhere, somehow, in early life, at the stage when children first acquire ritual responsiveness, Turing must have been busy with something else.

Brute-force Computation

The Robinson Crusoe quality was only one part of it. Not only independence of received knowledge, but avoidance of received *styles* (whether implanted by fashion or by long tradition) gave him a form of pleasure not

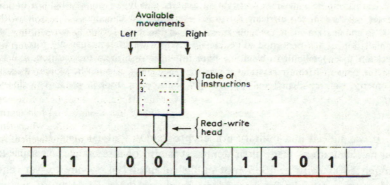

FIGURE 3.1   *Constituents of a Turing machine.* If a new 'table of instructions' is supplied for each computation, then each use creates a new, special-purpose, machine. If a once-and-for-all table ('language') is supplied, so that the specification of any given special machine which it is to simulate are placed on the input tape, then we have a universal Turing machine.

unmixed with glee. There was much of this in his recurrent obsession with attacking deep combinatorial problems by brute-force computation. This was at the heart of some of his cryptanalytical successes — notably his crucial inroad into the German Enigma cipher while working at Bletchley Park. It is difficult now to remember how startling, and to persons of mathematical taste how grating and offensive, was the notion of near-exhaustive enumeration of cases as an approach to a serious problem. Yet some recent negative reactions to Ken Appel and Wolfgang Haken's computer-aided proof of the four colour theorem gives a base from which to extrapolate back to the year 1943, the year my personal acquaintance with Alan Turing was formed.

Of course the abstract notion of combinational exhaustion was already deeply entrenched in mathematics. But what about the use of a physical device to *do* it? To make such proposals in earnest seemed to some people equivalent to bedaubing the mathematical sub-culture's precious tapestry with squirtings from an engineer's oilcan. Writing of an earlier juncture of intellectual history, Plutarch has left an unforgettable account:

Eudoxus and Archylas had been the first originators of this far-famed and highly prized art of mechanics, which they employed as an elegant illustration of geometrical truths, and as a means of sustaining experimentally, to the satisfaction of the senses, conclusions too intricate for proof by words and diagrams ... But what with Plato's indignation at it, and his invectives against it as the mere corruption and annihilation of the one good of geometry – which was thus shamefully turning its back on the unembodied objects of pure intelligence to recur to sensation, and to ask for help . . . from matter; so it was that mechanics came to be separated from geometry, and, repudiated and neglected by philosophers, took its place as a military art.

It was indeed in a military art, cryptography, that Turing's first practical mechanisations made their debut. It is also of interest that in a paper submitted as early as 1939 (not published until 1943 owing to war-time delays) a mechanisable method is given for the calculation of Georg Riemann's zeta-function suitable for values in a range not well covered by previous work. Why was Turing so interested? The answer would undoubtedly serve as another red rag to Plato's ghost, for the point at issue was a famous conjecture in classical pure mathematics: do all the zeros of the Riemann function lie on the real line? In a post-war paper the oil-can reappears in an attempt to calculate a sufficiency of cases on a computing machine to have a good chance either of finding a counter-example and thus refuting the Riemann hypothesis, or alternatively of providing non-trivial inductive support. The attempt, which was reported in the 1953 *Procceedings of the London Mathematical Society*, failed owing to machine trouble.

Machine trouble! Alan's robust mechanical ineptness coupled with insistence that anything needed could be done from first principles was to pip many a practical project at the post. He loved the struggle to do the engineering and extemporisation himself. Whether it all worked in the end sometimes seemed secondary. I was recruited at one point to help in recovering after the war some silver he had buried as precaution against liquidation of bank accounts in the event of a successful German invasion. After the first dig, which ended in fiasco, we decided that a metal detector was needed. Naturally Alan insisted on designing one, and then building it himself. I remember the sinking of my spirits when I saw the contraption, and then our hilarity when it actually seemed to be working. Alas its range was too restricted for the depth at which the silver lay, so that positive discovery was limited to the extraordinary abundance of metal refuse which lies, so we found, superficially buried in English woodlands.

The game of chess offered a case of some piquancy for challenging with irreverent shows of force the mastery which rests on traditional knowledge. At Bletchley Park, Turing was surrounded by chess-masters who did not scruple to inflict their skill upon him. The former British champion Harry Golombek recalls an occasion when instead of accepting Turing's resignation he suggested that they turn the board round and let him see what he could do with Turing's shattered position. He had no difficulty in winning. Programming a machine for chess played a central part in the structure of Turing's thinking about broader problems of artificial intelligence. In this he showed uncanny insight. As a laboratory system for experimental work chess remains unsurpassed. But there was present also, I can personally vouch, a Turing streak of iconoclasm: what would people say if a machine beat a master? How excited he would be today when computer programs based on his essential design are regularly beating masters at lightning chess, and producing occasional upsets at tournament tempo!

Naturally Turing also had to build a chess program (a 'paper machine' as he called it). At one stage he and I were responsible for hand-simulating and recording the respective operations of a Turing-Champernowne and a Michie-Wylie paper machine pitted against each other. Fiasco again! We both proved too inefficient and forgetful. Once more Alan decided to go it alone, this time by programming the Ferranti Mark 1 computer to simulate both. His problems, though, were now compounded by 'people problems', in that he was not at all sure whether Tom Kilburn and others in the Manchester laboratory, where he was working at the time, really approved of this use for their newly hatched prototype. It was characteristic of Turing, who was in principle anarchistically opposed to the concept of authority or even of seniority, that its flesh-and-blood realisations tended to perplex him greatly. Rather than confront the matter directly, he preferred tacitly to confine himself to nocturnal use of the machine. One way and another, the program was not completed.

It is fashionable (perhaps traditional, so deep are sub-cultural roots) to pooh-pooh the search-oriented nature of Turing's thoughts about chess. In his Royal Society obituary memoir, Max Newman observes in words of some restraint that '. . . it is possible that Turing under-estimated the gap that separates combinatory from position play'. Few yet appreciate that by setting the ability of the computer program to search deeply along one

FIGURE 3.2   The paradigm, derived by Turing and Claude Shannon for game-playing, implemented on an IBM 3-million-instructions-per-second computer, probes beyond the tactical horizons of even a grandmaster. In this match from Toronto in 1977, 'Kaissa', playing black, continued R-K1. It looks like a blunder — but was it?

line of attack on a problem in concert with the human ability to conceptualise the problem as a whole, programmers have already begun to generate results of deep interest. I have not space to follow the point here, but will simply exhibit, in Figure 3.2, a paradigm case. Here a program cast in the Turing-Shannon mould, playing another computer in 1977, apparently blundered. The chess-masters present, including former world champion Mikhail Botvinnik, unanimously thought so. But retrospective analysis showed that in an impeccably pure sense the move was not a blunder but a brilliancy, because an otherwise inescapable mate in five (opaque to the watching masters) could by this sacrifice be fended off for another 15 or more moves.

The equivocal move by Black, who has just been placed in check by the White Queen in the position shown, was 34 . . . *R-K1*, making a free gift of the Rook. The program, Kaissa, had spotted that the 'obvious' 34 . . . *K-N2* could be punished by the following sequence:

35.	*Q-B8* ch.	*K* x *Q* (forced)
36.	*B-R6* ch.	*B-N2* (or *K-N* 1)
37.	*R-B8* ch.	*Q-Q1*
38.	*R* x *Q* ch.	*R-K1*

39. *R* x *R* mate.

Suppose now that we interpret the situations-and-actions world of chess as a parable of computer-aided air-traffic control, or regulation of oil platforms or of nuclear power stations. If assigned to monitoring duty, Grandmaster Botvinnik would undoubtedly have presumed a system malfunction and would have intervened with manual override! Kaissa's deep delaying move (in the parable affording respite in which to summon ambulances, fire engines, etc.) would have been nullified.

Broader Horizons

Increasing numbers of industrial and military installations are controlled by problem-solving computing systems. The cloak cast by combinatorial complexity over the question of machine malfunction has thus acquired topical urgency. Only from computer analyses of chess along lines first charted by Alan Turing are documented and experimentally manipulable instances of the phenomenon yet available.

Turing's post-war plunge into mathematical models of embryo growth cast him once more in his favourite role as *enfant terrible* in virgin territory. At that time I was doing a rather stodgy genetics DPhil at Oxford. We corresponded about his new pursuit. Turing wanted to know whether this or that assumption underlying his formulations were biologically reasonable. Despite my relatively tender age I found myself carping at Turing's evident and self-confessed ignorance of genetics and embryology. In the process I managed to miss the point. Turing's contribution sustained an active line of theoretical biology, pushed ahead today by John Maynard Smith among others.

The A.M. Turing Trust, formed some seven years ago, has assisted in the collection, preparation and preservation of a variety of scientific and other records of his life. An appeal has now been launched to endow a Turing lecture, to be delivered annually at a British centre of learning – thus commemorating a distinguished Englishman and promoting the several fields of scientific inquiry which he was responsible for opening.

CHAPTER 4

Bayes, Turing and the Logic of Corroboration

Everybody knows about Bayes' theorem, which turns ordinary probability on its head; everybody, that is, except machine intelligence people. (But see a recent note by Duda, Hart and Nilsson, 1976.) This is curious since it was Alan Turing who as early as 1940 transformed into something computationally tractable Bayes' horrendous expression:

$$p(H/E) = \frac{p(H) \times p(E/H)}{p(E)}$$

which says that the (posterior) probability that a hypothesis H is true given an observed event E is the prior probability of H times the probability of E given H, all divided by the probability of E. This last quantity is computed as $p(H) \times p(E/H) + p(\bar{H}) \times p(E/\bar{H})$, where $\bar{H}$ means not $-H$.

The expression is 'horrendous' because a requirement to transform prior into posterior degrees of belief in H typically presents itself in a context of iteration:

$$\begin{array}{ccccc} p_0 \to p_1 \to p_3 \to & \ldots & \to p_n. \\ E_1 \quad E_2 & & E_n \end{array}$$

Here the degree of belief, p, is incremented and decremented by successive up-dates as first E_1, then E_2, then E_3 . . . etc. become available. Yet Bayes' original expression does not conveniently lend itself to incremental up-dating in this style, and the calculations are rather messy. Consider a very simple example.

A pot of 85 pennies is known to contain one double-headed penny. The rest are normal. A penny is sampled from the pot at random, is tossed 7 times, and is observed to fall 'heads' each time. Initially $p(H)$ was $1/85$. We wish to calculate $p(H/E_1)$, where E_1 is the event of 'all of 7 tosses of the coin gave "heads" '.

Applying the earlier expression we have:

$$P(H/E_1) = \frac{\dfrac{1}{85} \times 1}{\dfrac{1}{85} + \dfrac{84}{85} \times \dfrac{1}{128}} = p_1$$

which evaluates to approximately 3/5. Now we throw the coin once more and get another 'head'. The updating expression is:

$$p(H/E_1 \wedge E_2) = \frac{p_1 \times 1}{p_1 + (1 - p_1) \times \frac{1}{2}} = p_2$$

substituting the numerical expression above for p_1 in the lower expression already begins to generate unpleasant arithmetic. Rather than pursue elaborations, this note draws attention to the existence of a neat alternative.

The way in which Turing cut the computation knot (see Chapter 6 of I.J. Good's *Probability and the Weighing of Evidence*, 1950) was to say:

1) convert everything from probabilities into odds

i.e. $\dfrac{p}{1-p}$;

2) having done that, convert everything to logarithms.

Writing $U(H)$ to mean the prior odds in favour of some hypothesis H, and $U(H/E)$ for the posterior odds in favour of H given some evidence E, the following beautiful consequence follows from Bayes' original formulation:

$$\log U(H/E) = \log U(H) + w_1 + w_2 + \ldots + w_i + \ldots w_n$$

where w_i ('weight of evidence') is:

$$w_i = \log f_i = \log \frac{p(E_i/H)}{p(E_i/\bar{H})};$$

f_i is sometimes called 'the Bayesian factor in favour of H yielded by E_i': this was in fact Turing's terminology. $\bar{H}$ means the proposition that H is not true. So as each new E_i comes along we simply add in the corresponding w_i (which may be positive or negative) and update our log odds! Nothing could be more incremental, or more computationally simple.

It becomes possible with great ease to render various gnats strained at by philosophers into a pleasing Bayesian soup. Consider ravens. The assertion "All ravens are black" written:

1) $\forall x.$ raven $(x) \Rightarrow$ black (x)

is logically equivalent to

2) $\forall x.$ not (black $(x)) \Rightarrow$ not (raven (x))

We make an observation which leads us to assert additionally:

$\exists y.$ not (black (y)) $\wedge$ not (raven (y))

e.g., we have seen a green vase. Why does it not 'corroborate' 2) in the same way that a black raven 'corroborates' 1)?

Well actually it does, but not in the same degree! Enter Bayes.

We first have to fix up an interpretation of just what model is entailed in H and in not-H. For example we can express the definition

$H \equiv \forall x.$ raven $(x) \Rightarrow$ black (x)

as a frequency Table as follows:

	Black	Non-black		
Ravens	p	0	p	I here use the convention
Non-ravens	$p' - p$	q'	q	$q = 1 - p$
	p'	q'	1	$q' = 1 - p'$.

p is the frequency of ravens among all observable objects in the world. p' is the frequency of black objects among all the observable objects in the world.

As for the negation of H, a reasonable explication might be:

$\bar{H} \equiv$ (no correlation, as in the following Table):

	Black	Non-black		
Ravens	$p \times p'$	$p \times q'$	p	
Non-ravens	$q \times p'$	$q \times q'$	q	$q = 1 - p$
	p'	q'	1	$q' = 1 - p'$

Multiplication is the appropriate operation for mutually independent probabilities, if we wish to calculate the probability of the joint condition: thus the probability that x is the black raven is equal (assuming independence, i.e., non-correlation) to the probability that x is a raven times the probability that x is black.

Now let

E_i = 'on my i-th experiment I catch a raven and I observe that it is black'

To calculate f we take the ratio of the top left cell of the first Table to that of the second Table:

$$f = \frac{p(E_i/H)}{p(E_i/\bar{H})} = \frac{p}{pp'} = \frac{1}{p'}$$

Note that f is large and hence that $\log f$ is *large* (because only a small minority of things are black). But contrast

E_j = 'on my j-th experiment I find a green (non-black) vase (non-raven)'

This time we take the ratio of the bottom right cell of the first Table to that of the second Table:

$$f = \frac{p(E_j/H)}{p(E_j/\bar{H})} = \frac{q'}{qq'} = \frac{1}{q}$$

Note that f is just a little greater than 1 and hence that $\log f$ is *small* — near zero — (because most objects in the world are non-ravens — i.e. ravens are rare).

Naturally everday mental practice (perfectly sound) is to disregard *small* weights of evidence — i.e. to perceive them as having zero weight. Hence the apparent corroboration paradox about ravens. But although the true weight in favour of $\bar{H}$ is non-zero the weights *are* unsymmetrical between the two inferences; only quantitatively, not qualitatively as the term 'paradox' would imply.

There can be practical consequences. For example, for 'ravens' read 'epileptics' and for 'are black' read 'have fits'. We would normally disregard the corroboration yielded by the discovery that some fit-free person does not have epilepsy. But what if we narrow our definition of 'the world'? What if the discovery is made in an epilepsy clinic? *Then* the corroboration will be extremely strong! P. Suppes (1966), who was the first to note the relevance of Bayes to the raven paradox, makes essentially the same point when he proposes using the inequality of $q' > p$ (our notation) as the necessary condition for the 'paradox' to look like one.

REFERENCES

Duda, R.O., Hart, P.E. and Nilsson, N.J. (1976). Subjective Bayesian methods for
    rule-based inference systems. *Technical Note 124*, Stanford Research Institute.
Good, I.J. (1950). *Probability and the Weighing of Evidence*. London: Charles
    Griffin.
Suppes, P. (1966). A Bayesian approach to the paradoxes of confirmation. *In Aspects
    of Inductive Logic*, Amsterdam: North-Holland.
In addition Shortliffe, E.H. and Buchanan, B.G. (1975) (A model of inexact
reasoning in medicine, *Mathem. Biosciences*, **23**, pp. 351–379) have an interesting
proposal for handling incremental corroboration in the case when sufficient statistical
data is not available to underpin a strict Bayesian calculus.

Introductory Note to Chapter 5

Representation of problem-solving as a search for one or more of a set of 'goal' nodes of a directed graph loomed large in the first 20 years of attempts on the mechanisation of thought processes. At times the pursuit seemed forlorn. Awareness that heuristic search is not enough was paralleled by a lack of clear ideas as to what forms the rest of the needed apparatus might take.

New 'advice languages' for computers with inbuilt facilities for deductive and inductive logic are now pointing the way. Heuristic search has meanwhile attained the honorable status of a cliché, much as in a mechanical device pistons and wheels are not spoken about but assumed. This article is based on a lecture delivered to the Edinburgh University Mathematical Society which gave an introduction to the theory and surveyed the state of the art.

CHAPTER 5
Heuristic Search

In any rapidly expanding new technology, and machine intelligence is such a technology, three phases can commonly be identified. Typically they overlap or merge with each other in time.

First comes the phase of *ad hoc* innovation, exemplified by the Wright brothers in aeronautics or Watt in the development of steam power. In addition to their practical skills as engineers or experimentalists, the pioneers may require to bring to the attempt new instruments and techniques imported from other fields, native ingenuity, persistence, and the luck or foresight to have selected a task which is actually possible of accomplishment with the materials and methods available in their generation. But the one ingredient which is usually lacking at this primitive moment of creation is a theoretical framework adequate to enable these workers, or anyone else, to understand properly what they are doing. In the long run, of course, this ingredient is the crucial one.

Next comes the phase of formalisation. Enter the mathematicians, their curiosity provoked by the spectacle of unexplained achievements and clumsy contrivances lacking perceptible principles of rational design. This is the phase in which a new science is born, thermodynamics in the wake of the steam engineers, aerodynamics in the wake of the fliers, information theory and switching theory in the wake of Marconi and Bell.

The final phase is one of stabilisation and systematic development. The mathematician still holds the commanding position, but the type of mathematician has changed. The research mathematicians have moved on to virgin fields, either of abstraction or of newer application areas. In their place are practitioners; to a large extent they may be engineers who are required in this third phase to have a standard repertoire of mathematical technique in order to be professionally qualified at all. This is the sign that the new technology has attained full maturity, and that it is time for any mathematically unversed survivors from the first phase to get with it or get out.

Machine intelligence as a whole is today in headlong transit from phase 1 to phase 2. I derive particular interest from this opportunity of talking to students of one of the most rapidly developing mathematical schools in the country; because if there was ever a moment when a fast-moving new technology needed young mathematical blood, and could offer the the excitement of challenging but essentially tractable problems, that moment is now. I should add, in a more personal vein, that I am not myself a mathematician, although I have spent several stretches of my life working with mathematicians — Newman, Good, Whitehead, Turing and others during the war, with R.A. Fisher during the early 1950s, and in recent years with my own present research colleagues. I am now engaged in a process which I am finding thrilling, daunting, absorbing and nerve-racking by turns — the process of acquiring sufficient elements of relevant parts of mathematics to make sense of what I myself and many earlier workers have been doing in one specialised sector of the research front, usually known as 'heuristic search'.

Problems and Graphs

A class of problem which includes those problems which are susceptible to heuristic search can be defined, following Ernst and Newell (1969), in the following terms:

Given: an initial state
         a definition of a desired state
         a repertoire of actions,
Find: a sequence of actions which will transform the initial state into
         the desired state.

Notice that the above formulation defines the class of one-person games. The approach to mechanising the search for solutions in such situations is through a graph-theoretical representation as follows:

Given: a node of a connected graph
         a set of nodes satisfying a goal predicate
         a set of operators,
Find: a sequence of operator-applications which will generate a path

leading from the initial node to a goal node.

This mapping of problems onto graphs is in essence an old one. In recent times it has been developed by von Neumann and Morgenstern (1944) and used in early studies of computer game-playing by Turing (1953), Shannon (1950), Strachey (1952) and above all Samuel (1959, 1967), whose study of machine learning using the game of checkers is a notable example of the kind of *ad hoc* pioneering development which I discussed earlier. Samuel was the author of the first computer program capable of defeating the general run of amateur players at a non-trivial game (I define a non-trivial game as one for which world championship matches occur). His achievement, within its limited frame of reference, is comparable with the inventions of Watt and Stephenson. It also resembles theirs in being rich in potential theory, but lacking the formal apparatus for describing the artefact. This article is largely concerned with the progress which has subsequently been made towards the construction of a descriptive apparatus of the required kind.

A number of features in this progression can be identified:

1) Adaptation by Newell, Shaw and Simon (1957), Newell and Simon (1961), of the graph representation to the case of one-person games in the form a series of computer programs for 'general problem-solving', known collectively as GPS. These authors established a clear separation between the general part (graph-searching program) and the task-specific part (application package).

2) Design and formal description by Doran and Michie (1966) of the Graph Traverser algorithm embodying the lookahead and evaluation procedures of classical game-playing. Hart, Nilsson, and Raphael (1967) developed a variant form of essentially the same algorithm which allowed for variable costs associated with different operator-applications.

3) A fundamental theorem (Hart, Nilsson, and Raphael *loc. cit.*) which states conditions under which the benefits of heuristic search (saving of computational work) can be got without sacrifice of the minimality of the cost associated with the final solution path (calculated as the sum of the costs of the operator-applications required to generate the path). Previously the only known guarantee was by exhaustive breadth-first search (Moore, 1959; Dantzig, 1963; Dijkstra, 1959).

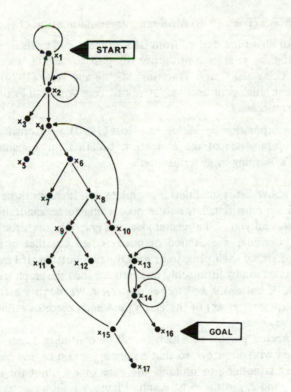

Nodes	Arcs from each node	Nodes	Arcs from each node
x_1	$e_{1,1}; e_{1,2};$	x_{10}	$e_{10,4}; e_{10,13}$
x_2	$e_{2,1}; e_{2,2}; e_{2,3}; e_{2,4};$	x_{11}	–
x_3	–	x_{12}	–
x_4	$e_{4,5}; e_{4,6};$	x_{13}	$e_{13,13}; e_{13,14};$
x_5	–	x_{14}	$e_{14,13}; e_{14,14}; e_{14,15}; e_{14,16};$
x_6	$e_{6,7}; e_{6,8};$	x_{15}	$e_{15,11}; e_{15,17};$
x_7	–	x_{16}	–
x_8	$e_{8,9}; e_{8,10};$	x_{17}	–
x_9	$e_{9,11}; e_{9,12};$		

FIGURE 5.1. A directed graph, with 'start' and 'goal' nodes.

4) Proofs (Pohl, 1970) concerning the minimisation of search cost.

5) An algorithm derived from GPS which re-orders its operators, on the basis of their relative contribution to past successful searches (Quinlan, 1969). Using the Graph Traverser, Michie and Ross (1970) have demonstrated an analogous method of re-ordering which does not depend on previous successes.

6) Incorporation by Michie and Ross (1970) of automatic optimisation of the parameters of the evaluation function, in a manner similar to Samuel's 'learning by generalisation'.

The above list is doubtless incomplete, but includes those features with which I am most familiar. Before proceeding to an annotation of the list, I shall remind you of the general idea of a graph, in the sense used in graph theory. Consider a collection of buttons tied together in some way by pieces of string; each piece joins exactly two buttons. If I can pick up the entire collection by lifting only one button, then the graph representing my network of buttons is said to be *connected*. We identify each button with a *node* (point, vertex) of the graph, and each piece of string with an *arc* (line, edge).

In *directed* graphs, to which I shall be confining discussion, every arc is marked with an arrow, so that an arc $e_{i,j}$ is said to lead from node x_i to x_j. There is nothing to prohibit the case of $i = j$, nor the occurrence of both $e_{i,j}$ and $e_{j,i}$ in the same graph: Figure 5.1 shows a very simple graph which exhibits both features.

A natural notation for describing this graph would be an expression in terms of

1) A set of nodes $\{x_1, x_2, \ldots, x_{17}\}$, which we may denote X, and

2) A set of arcs $\{e_{1,1}, e_{1,2}, \ldots, e_{15,17}\}$, which we may denote E.

More exactly, we could say that a graph is an ordered pair (X, E) of which the first element is the set X and the second is the set E. An alternative approach, more convenient to our purpose, is to define the set E not explicitly, as above, but implicitly by introducing a function Γ, which maps from each given node to the set of its successors, as illustrated in Table 5.1. Given X and Γ it is clearly possible to generate the entire graph, which may thus be represented by the ordered pair (X, Γ).

TABLE 5.1
The function Γ defined over the graph of Figure 5.1.

ARGUMENT	VALUE
x_1	$\{x_1, x_2\}$
x_2	$\{x_1, x_2, x_3, x_4\}$
x_3	$\{\ \}$
x_4	$\{x_5, x_6\}$
x_5	$\{\ \}$
x_6	$\{x_7, x_8\}$
x_7	$\{\ \}$
x_8	$\{x_9, x_{10}\}$
x_9	$\{x_{11}, x_{12}\}$
x_{10}	$\{x_{13}\}$
x_{11}	$\{\ \}$
x_{12}	$\{\ \}$
x_{13}	$\{x_{13}, x_{14}\}$
x_{14}	$\{x_{13}, x_{14}, x_{15}, x_{16}\}$
x_{15}	$\{x_{11}, x_{17}\}$
x_{16}	$\{\ \}$
x_{17}	$\{\ \}$

We now consider the use of graphs to represent problems. To be susceptible of a conventional graph representation a problem must take the form of a set of discrete states, one of which is designated the initial state, together with a repertoire of actions. An action operates on an individual state to transform it into a derived state. Each action is thus conceived as a unary function mapping the set of states into itself. A solution consists of a chain of actions which can successively convert the initial state into a goal state, specified either explicitly or by a goal-recognising rule. In the representation, states of the problem are identified with the nodes of the graph, and actions are identified with unary operators. Each operator is defined over some subset of X, not the whole set because an action is typically only applicable to a subset of the states of the problem (e.g. P-K4 is a legal move in some, but not all, Chess positions). The operator corresponding to a given action is thus a partial function over X, yielding the value *undefined* when applied outside the appropriate subset of X. Writing $\{\Gamma_1', \Gamma_2', \ldots, \Gamma_m'\}$ to denote the set of operators, we could exhibit the relationship to the function Γ by attaching labels to the arcs

TABLE 5.2

An interpretation of the arcs shown in Figure 5.1 in terms of a set of operators
$\Gamma = \{\Gamma_1{}', \Gamma_2{}', \Gamma_3{}', \Gamma_4{}'\}$. In the key, these operators are related to a repertoire of
actions (see text).

Arcs of the Graph	Operators
$e_{1,1}$	$\Gamma_2{}'$
$e_{1,2}$	$\Gamma_1{}'$
$e_{2,1}$	$\Gamma_1{}'$
$e_{2,2}$	$\Gamma_2{}'$
$e_{2,3}$	$\Gamma_4{}'$
$e_{2,4}$	$\Gamma_3{}'$
$e_{4,5}$	$\Gamma_3{}'$
$e_{4,6}$	$\Gamma_4{}'$
$e_{6,7}$	$\Gamma_3{}'$
$e_{6,8}$	$\Gamma_4{}'$
$e_{8,9}$	$\Gamma_3{}'$
$e_{8,10}$	$\Gamma_4{}'$
$e_{9,11}$	$\Gamma_3{}'$
$e_{9,12}$	$\Gamma_4{}'$
$e_{10,4}$	$\Gamma_4{}'$
$e_{10,13}$	$\Gamma_3{}'$
$e_{13,13}$	$\Gamma_2{}'$
$e_{13,14}$	$\Gamma_1{}'$
$e_{14,13}$	$\Gamma_1{}'$
$e_{14,14}$	$\Gamma_2{}'$
$e_{14,15}$	$\Gamma_3{}'$
$e_{14,16}$	$\Gamma_4{}'$
$e_{15,11}$	$\Gamma_4{}'$
$e_{15,17}$	$\Gamma_3{}'$

KEY:	*Operator*	*Action*
	$\Gamma_1{}'$	Rotate 180°
	$\Gamma_2{}'$	Rotate 360°
	$\Gamma_3{}'$	Drive forward
	$\Gamma_4{}'$	Turn right

of the graph of Figure 5.1. A graph so labelled is called a *coloured graph*,
because one way of labelling arcs would be by the use of coloured inks.
This is indicated in Table 5.2 which is based on a particular interpretation
of the graph in the form of a simple navigation problem. In this problem,
which is illustrated in Figure 5.2, an automobile lacking a reverse gear and

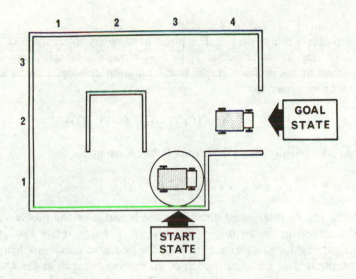

Repertoire of Actions: 1. **Rotate turntable 180°**

2. **Rotate turntable 360°**

3. **Drive car forward unit distance**

4. **Turn right**

N.B.   The turntable can only be operated when it has the weight of a car on it.

FIGURE 5.2.   A problem which maps onto the graph of Figure 5.1. Continuous double lines indicate walls. The automobile cannot reverse nor turn left. A sequence of actions is required, which will get it from the start state (3, 1, *East*) to the goal state (4, 2, *East*).

unable to turn left must be driven through a road system which includes (right-hand bottom corner) a turn-table capable of two movements: a rotation through 180° and a rotation through 360°. These account for two out of the total repertoire of four operators associated with the problem. The other two are: 'drive one unit straight forwards' and 'turn right' (the latter has the effect of 'one unit straight forwards, rotate clockwise 90°, then one more unit straight forwards'). A state of the problem takes a form such as 'automobile faces South on square (3, 2)'. This might be written as a triple: (3, 2, *South*). It is required to find a path from x_1 to

x_{16}, i.e. from $(3, 1, East)$ to $(4, 2, East)$.

The problem is a trivial one. It is posed here merely to clarify notation. As a final step in this direction Table 5.3 defines the complete operator set in terms of the problem graph. In the notation developed here a solution can be expressed as

$$x_{16} = \Gamma_4{}' \, (\Gamma_1{}'(\Gamma_3{}'(\Gamma_4{}'(\Gamma_4{}'(\Gamma_4{}'(\Gamma_3{}'(\Gamma_1{}'(x_1))))))))),$$

or more conveniently, using the reverse Polish convention, as

$$x_{16} = x_1 \, \Gamma_1{}' . \Gamma_3{}' . \Gamma_4{}' . \Gamma_4{}' . \Gamma_4{}' . \Gamma_3{}' . \Gamma_1{}' . \Gamma_4{}';$$

The latter can be interpreted directly in the language of the problem as a list of instructions to the driver: 'Start at $(3, 1, East)$, rotate $180°$, forward, right, right, right, forward . . . etc.' The heuristic search problem can thus be phrased as the construction of an expression, such as the above, which gives the goal in terms of the start. This can be seen to be equivalent to a search for a path across the graph.

Moore's Algorithm

The classical algorithm for finding the shortest path across a graph is due to Moore (1959). To expound it we represent a graph, G, as (X, Γ) where X is a set of nodes and $\Gamma : X \to 2^X$ gives the set of immediate successors of any node x in X to which it is applied. Given an initial node x_0 and a goal-recognising predicate P defined over X it is required by successive applications of Γ to find a path $x_0, x_1, \ldots, x_k$ such that $P(x_k)$ and k is a minimum.

Moore's algorithm, having labelled x_0 with the integer 0, applies Γ to it and labels each unlabelled $x \in \Gamma(x_0)$ with the integer 1. At each application of Γ a pointer is placed leading from each member of the successor set back to its parent. If none of these successors satisfies P then Γ is applied to each of the 1-nodes in turn, labelling all unlabelled new nodes with the integer 2. In the interests of efficiency the generation of a new node is inhibited if it is equal to a node already generated and labelled. The process is iterated until either no new unlabelled nodes can be produced (i.e. no path exists) or a node is produced satisfying P. In the latter

TABLE 5.3
Results of applying the operators of Γ' to the nodes of X, based on Table 5.2 and Figure 5.2. The symbol '*undef*' is used for '*undefined*'.

	Γ_1'	Γ_2'	Γ_3'	Γ_4'
x_1	x_2	x_1	undef	undef
x_2	x_1	x_2	x_4	x_3
x_3	undef	undef	undef	undef
x_4	undef	undef	x_5	x_6
x_5	undef	undef	undef	undef
x_6	undef	undef	x_7	x_8
x_7	undef	undef	undef	undef
x_8	undef	undef	x_9	x_{10}
x_9	undef	undef	x_{11}	x_{12}
x_{10}	undef	undef	x_{13}	x_4
x_{11}	undef	undef	undef	undef
x_{12}	undef	undef	undef	undef
x_{13}	x_{14}	x_{13}	undef	undef
x_{14}	x_{13}	x_{14}	x_{15}	x_{16}
x_{15}	undef	undef	x_{17}	x_{11}
x_{16}	undef	undef	undef	undef
x_{17}	undef	undef	undef	undef

case the label gives the length of the minimal path, which can be reconstructed by tracking back along the pointers from the goal node to the root node.

If the arrows on these pointers are reversed, they identify a sequence of operators Γ_a', Γ_b', Γ_c', ... etc. An operator is a function $\Gamma_i': X \to X$, and for our present purpose we take

$$\bigcup_{i=1}^{m} \Gamma_i'(x) = \Gamma(x)$$

for all x in X, where m is the number of operators in the complete set, Γ', of transformations which can be used to effect the transition from a node to an immediate successor. $\Gamma_i' = \Gamma_j'$ if and only if $\Gamma_i'(x) = \Gamma_j'(x)$ for all x in X. To allow for the existence of operators which may be inapplicable to some nodes, we add an additional node to X which has the value *undefined*.

We thus represent the path $x_0, x_1, \ldots, x_k$ in terms of a sequence of operator-applications, thus: $\Gamma_a'(x_0)$, $\Gamma_b'(\Gamma_a'(x_0))$, $\ldots$, $\Gamma_t'(\Gamma_s'(\ldots(\Gamma_b'(\Gamma_a'(x_0)))\ldots))$. On this basis it is natural to consider the different members of Γ' as varying in the cost of application, and more generally to associate a specific cost c_{ij} with each arc leading from x_i to x_j. c_{ij} may be identified with the cost of evaluating $\Gamma_a'(x_i)$ where x_j is the value of the expression. A shortest path algorithm appropriate to this form of the problem was proposed by Dijkstra (1959).

The Graph Traverser

Moore's algorithm is sometimes called 'breadth-first' search, and like Dijkstra's, is extremely expensive in computational work, prohibitively so for large graphs with high branching ratios. Heuristic search studies have in large part been devoted to adaptations of algorithms of the Moore type with the aim of greatly reducing the amount of computation required. The basis of such adaptations is the idea of guiding the search preferentially in the more promising directions with the aid of a *heuristic function*. This concept, essentially similar to the 'evaluation function' of classical game-playing studies, was first applied to the graph searching problem by Doran and Michie, although a similar notion is discernible in the 'differences' of GPS, which is not based on Moore's algorithm. Doran and Michie proposed the use of a function, which we shall here call h, which is an estimator of the shortest distance from the given node to the goal node. Their algorithm, the Graph Traverser, proceeds exactly as Moore's except that at each stage the next node chosen for 'development', i.e. application of Γ, is not the node with the lowest-valued integer label but the node with the lowest distance-to-goal estimate. The effect of this is to provide a means of exploiting any estimator which may be available in order to cut down the search cost, measured, for example, by the size of the union of all the successor-sets produced by applications of Γ during the search. In favourable cases this reduction can be very drastic, but a penalty is paid in the sacrifice of optimality: it can no longer be guaranteed in general that the final path will be the shortest, although in certain applications cheap sub-optimal solutions may be preferred to expensive optimal ones.

Preserving Minimality

Hart, Nilsson, and Raphael showed a means of obtaining some of the search-reduction benefits conferred by the use of a heuristic function yet preserving the minimality of the final path found.

These authors introduced a function f defined as $f(x) = g(x) + h(x)$ for $x \in X$ where $g(x)$ is the distance of x in the search tree from the initial node x_0 and h is defined as previously. Thus in the case where each arc bears a unit cost, $g(x)$ corresponds to the labels attached to the nodes by Moore's algorithm, and a Graph Traverser search using g in place of h reduces to Moore's algorithm.

The Hart-Nilsson-Raphael theorem states that use of f in place of h is guaranteed to produce the minimum path if, for all $x, h(x) \leqslant h^*(x)$ where h^* is the true distance to the goal. The authors give as an example the relation between the air distance and the road distance from one city to another over a map. If the air routes are straight lines, then h is always bounded above by h^* in the required fashion. An example from the class of sliding block puzzles considered by Doran and Michie would be the quantity

$$\sum_{i=1}^{n} p_i$$

where p_i is the 'city-block distance' of the ith piece from its destination.

The theorem holds in the general case where different arcs have different costs, but this case will not be further considered here.

Pohl's Results

It is convenient to re-express the above result using a weighting factor ω, as introduced by Pohl (1970), and write $f(x) = (1 - \omega) \times g(x) + \omega \times h(x)$. The theorem tells us for a given h how far we can increase ω before optimality is lost. Specifically, it states that $\omega/(1 - \omega) \times h(x) \leqslant h^*(x)$, for all x in X.

Among results of interest proved by Pohl are:

1) If h is perfect, i.e. always equal to h^*, then the search effort (number of applications of Γ) is minimised if $\omega \geqslant \frac{1}{2}$. The final path is minimal for any $\omega, 0 \leqslant \omega \leqslant 1$.

2) Pohl has also shown that to minimise search effort when h is subject to error it is not necessarily correct to set $\omega = 1$ as in the original Graph Traverser. Proof of this result has so far been obtained only where the problem graph takes the special form of an infinite tree. Experimental results with non-tree problem graphs lend some support to a generalisation. But it is by no means clear that the reasons why the residual influence of the g term is beneficial in these experiments are strictly related to those from which the proof was constructed. An informal and imprecise suggestion can, however, be ventured in terms of the occurrence in problem-solving lookaheads of blind alleys which can run to various depths. The lower the value of ω — i.e. the greater the weight given to the g term: 'How far have I come?' — the sooner is each blind-alley excursion curtailed, and the search process drawn back to more promising terrain.

3) Pohl (1971) has further shown that if some solution path exists, then a search with $\omega < 1$ will find one. The argument is that *all* nodes with $g = 1$ must eventually be expanded, even the one with the highest value of $f(x)$, say M. Clearly even this node will be developed before any node at distance more than $M/(1 - \omega)$ from the start because a node at this distance must have $f(x) > M$. The argument can be applied inductively to show that all nodes with $g = 2$, all nodes with $g = 3, \ldots$ etc. must be developed in the end.

Automatic Optimisation of Search Efficiency

Studies of automatic optimisation by Michie and Ross were conducted with a modified version of the Graph Traverser, GT4, incorporating a feature first implemented by Doran (1968) and referred to as 'partial development'. As in Doran's version, GT4 applies only one operator at a time from the set Γ'. The generation of the path $x_0, x_1, \ldots, x_k$ is performed by a function *transform* $: X \to S$ where S is the set of all paths from x_i to x_j where *not $P(x_i)$ and $P(x_j)$*. The evaluation of *transform* applied to some node x, proceeds as follows:

1) print (x)
2) if $P(x)$ then exit
3) assign *strategy* (x) to *operator*

4) assign *operator* (x) to x

5) go to step 1.

The function *strategy* : $X \to \Gamma'$ produces the next operator to apply to x, i.e. the next action to apply to the current state of the problem. It is in the evaluation of this function that the entire lookahead tree is grown, at each stage applying to that node on the tree for which h is a minimum, x_{min}, the first operator in the ordered set $\{\Gamma_1', \Gamma_2', \ldots, \Gamma_m'\}$ which has not previously been applied to it. When a preset limit to the size of the lookahead tree is reached, a path is traced from the current x_{min} back to the root node, and the member of Γ' which corresponds to the first arc of the retraced path is the result delivered by *strategy*.

GT4 is open at two points to the incorporation of 'learning' features.

1) The action of h can be optimised by automatic adjustment of a global list of parameters used in its evaluation. The regime adopted in tests of this idea bears similarities to the 'learning by generalisation' of Samuel's (1959) checkers-playing program. But it optimises a different measure of performance, more appropriate to graph searching; namely the accuracy of h as an estimator of distance over the stored search trees generated by *strategy* (*see* Doran, 1967). These can be regarded as local samples of the graph over which a good distance estimator is desired.

2) The set Γ' can be automatically re-ordered so as to bring operators of proven utility to the top of the list. The method adopted was to promote, on exit from *strategy*, the selected operator and to demote any rival operators which had been applied to the same root node during the evaluation of *strategy*.

Both the above self-optimising devices were found by Michie and Ross to give substantial improvements of search efficiency. But once again, improvement is purchased at the price of optimality of the final solution, since GT4 has outgrown the bounds within which the Hart-Nilsson-Raphael result was proved.

Applications

To ask for examples of the practical usefulness of heuristic search would at one time have been reminiscent of the question 'What use is a new-born baby?' Today it is more like asking 'What use is a teen-ager?' since the main feature now is the sheer multiplicity of potential applications, many of them as yet untried. The case is similar to that of automatic methods of minimising continuous functions of numerical variables, of which heuristic search can be regarded as a generalisation (see Michie and Ross, *loc. cit.*, for a worked example). If we were asked today 'What use are numerical minimisation methods?' we would probably reply with a roll call: physics, chemistry, crystallography, astronomy, aerodynamics, hydrodynamics, tribology, mechanical engineering, automatic control, etc. etc. With heuristic search (i.e. minimisation in non-dimensional, discrete, spaces) we are beginning to build up the elements of a similar roll call: game-playing problems, shortest route problems, allocation problems, distribution problems, symbol-manipulation problems, scheduling problems, recognition problems, etc. etc. To convey the flavour of applications which give an opening to heuristic search methods I shall describe two examples – a distribution problem, and a symbol-manipulation problem from artificial intelligence research.

The distribution problem, adapted from Nilsson (1971) is on a toy scale. But it is easy to envisage extensions and complications such as would be encountered in real life, were it desired to design by computer a large-scale lay-out of pipes so as to distribute fluid from a number of sources to a number of sinks under constraints (compare the electricity distribution problem of Burstall, 1966). In the example shown in Figure 6.3, two fluid sources, A, with a capacity of 4 gallons/minute and B, with a capacity of 2 gallons/minute, must supply two sinks C and D with requirements of 3 gallons/minute each, through pipes of maximum transmission capacity of 3 gallons/minute. If the sources and sinks are located as in the figure, and if we allow pipe connections only at source or sink locations, how should the pipes be laid so as to use the least amount of pipe?

States can be represented as 4-tuples: thus the initial state, A = 4, B = 2, C = 0, D = 0, is represented as (4, 2, 0, 0), and the goal state as (0, 0, 3, 3).

Actions correspond to transferring the increment of 'fluid per minute' from one point to another. We will also allow 'compound actions', which transfer two or three increments in a single step. The corresponding operators, 36 in number, are:

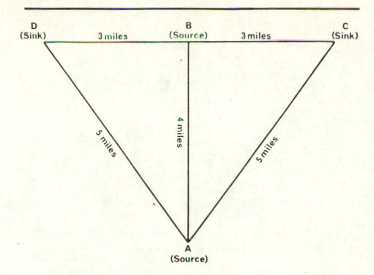

FIGURE 5.3.   Locations of fluid sources and sinks (see text).

Operator   *Action*
AB,(AB)², (AB)³ transfer 1, 2, or 3 gal./minute from A to B
AC,(AC)², (AC)³ transfer 1, 2, or 3 gal./minute from A to C

.
.
.
.

.

BA,(BA)², (BA)³ transfer 1, 2, or 3 gal./minute from B to A

An action is only applicable if there is sufficient extra fluid at the point from which fluid is to be transferred, and we shall also exclude actions which overshoot a sink's requirement, or which directly reverse earlier actions. Each transfer may commit additional pipe to effect it.

A fragment of the heuristic search tree is shown in Figure 5.4. Each state representation is preceded for convenience by a cost component, i.e. the total length of pipe characterising the given state. In the lower half of each box a value has been assigned computed from an evaluation function following the Hart-Nilsson-Raphael form $f = g + h$; in this case g

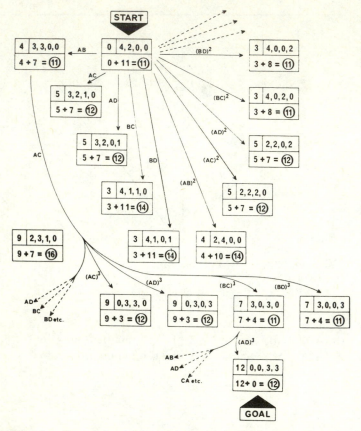

FIGURE 5.4.   Part of the search tree for a simple distribution problem (see text). Search is guided by an evaluation function of the form $f = g + h$, the value of which is given by the ringed figure in the bottom half of each box. A minimum-cost solution is guaranteed, but the search effort is reduced by employment of the heuristic term h.

is given by the pipe-length term which measures the cost of the steps taken so far, and for h I have taken $4 \times ((A/3)) + 3 \times ((B/3))$ where $((x))$ means 'x rounded up to the nearest integer'. This quantity is a lower bound on the amount of piping needed to relieve both sources, A and B, of their loads and hence has the required property of being everywhere bounded above by the true remaining cost-to-goal.

The second example, taken from an artificial intelligence research project (automation of inductive reasoning), is due to Popplestone (1970). The program is set the task of guessing the winning rule for tic-tac-toe (i.e. all of the squares of a row, or of a column, or of a diagonal of the 3 x 3 array must be occupied) from an incremental set of positive and negative instances. A 'guess' is a sentence of the predicate calculus formed by operating on a *current base set* of sentences by certain given rules. Initially the current base set consists only of the set of primitive sentences, such as *occ* (1, 2, '*X*'), meaning 'square 1, 2 is occupied by an *X*.' As the guessing game proceeds the rest of sentences is enriched by derived sentences, a process which is carried out by the Graph Traverser running as a sub-routine of the main induction program. The predicate calculus sentences correspond to nodes of the graph, while the formation rules correspond to the operators. The evaluation function expresses (1) how well the sentence agrees with the evidence of the sample tic-tac-toe boards collected so far, and (2) the complexity of the sentence: following William of Occam, the program gives preference to simplicity. A full account of the work, together with experimental results, is given in the original paper.

Future Developments

In the future, heuristic search studies, at least in the context of artificial intelligence, are likely to pay increased attention to devising automatic methods of imposing classification on problem spaces. The object is to sub-divide the space into regions, with each of which a different operator-selection regime can be associated. This is the spirit of GPS 'operator-difference' tables, but systematic methods of generating the required differences (features) continue to elude workers in the field. Closely allied to this is the search for automatic methods of finding improved representations of problems. These topics take us beyond the limits of heuristic search as strictly defined.

REFERENCES

Burstall, R.M. (1966). Computer design for electricity supply networks by a heuristic method, *The Computer Journal*, Vol. 9, pp. 263-274.
Dantzig, G. (1963). *Linear Programming and Extensions*, Princeton: Princeton

University Press.

Dijkstra, E. (1959). A note on two problems in connection with graphs, *Numerische Mathematik*, Vol. 1, pp. 269-271.

Doran, J.E., and Michie, D. (1966). Experiments with the Graph Traverser program, *Proc. R. Soc. (A)*, Vol. 294, pp. 235-259.

Doran, J.E. (1967). An approach to automatic problem-solving, *Machine Intelligence 1* (eds. N.L. Collins and D. Michie), Edinburgh: Oliver and Boyd, pp. 105-123.

Doran, J.E. (1968). New developments of the Graph Traverser, *Machine Intelligence 2* (eds. E. Dale and D. Michie), Edinburgh: Oliver and Boyd, pp. 119-135.

Ernst, G.W., and Newell, A. (1969). *GPS: A Case Study in Generality and Problem Solving*, New York and London: Academic Press.

Hart, P., Nilsson, N., and Raphael, B. (1967). A formal basis for the heuristic determination of minimum cost paths, *Stanford Research Institute Report*. Reprinted in *IEEE Trans. on Sys. Sci. and Cybernetics* (1968).

Michie, D., and Ross, R. (1970). Experiments with the adaptive Graph Traverser, *Machine Intelligence 5* (eds. B. Meltzer and D. Michie), Edinburgh: Edinburgh University Press, pp. 301-318.

Moore, E. (1959). The shortest path through a maze, *Proceedings of an International Symposium on the theory of switching. Part II*, pp. 285-292, Cambridge: Harvard University Press.

Newell, A., Shaw, J.C., and Simon, H.A. (1957). Preliminary Description of General Problem Solving Program – 1 (GPS-1), *CIP Working Paper No. 7*, Pittsburgh: Carnegie Institute of Technology.

Newell, A., and Simon, H.A. (1961). GPS, a program that simulates human thought, *Lernende Automaten*, (ed. H. Billing), p. 109. Reprinted in *Computers and Thought* (eds. E.A. Feigenbaum and J. Feldman) New York: McGraw-Hill (1963).

Nilsson, N. (1971). *Problem-solving methods in artificial intelligence*, New York: McGraw-Hill.

Pohl, I. (1970). First results on the effect of error in heuristic search, *Machine Intelligence 5* (eds. B. Meltzer and D. Michie), Edinburgh: Edinburgh University Press, pp. 219-236.

Pohl, I. (1971). Heuristic search viewed as path finding in a graph. *Artificial Intelligence*, Vol. 1, pp. 193-204.

Popplestone, R.J. (1970). In *Machine Intelligence 5* (eds. B. Meltzer and D. Michie), pp. 203-215.

Quinlan, J.R. (1969). A task-independent experience-gathering scheme for a problem solver, *Proc. International Joint Conference on Artificial Intelligence* (eds. D.E. Walker and L.M. Norton), p. 193.

Samuel, A.L. (1959). Some studies in machine learning using the game of checkers, *IBM. J. Res. Dev.*, Vol 3, pp. 211-229.

Samuel, A.L. (1967). Some studies in machine learning using the game of checkers, 2 – recent progress, *IBM J. Res. Dev.*, Vol. 11, pp. 601-617.

Shannon, C.E. (1950). Programming a computer for playing chess, *Phil. Mag.*, Vol.

41, p. 256.

Strachey, C.S. (1952). Logical or non-mathematical programmes, *Proc. Assoc. for Computing Machinery Meeting*, p. 46, Toronto.

Turing, A.M. (1953). Digital computers applied to games, *Faster than Thought*, p. 286 (ed. B.V. Bowden), London: Pitmans.

Von Neuman, J., and Morgenstern, O. (1944). *Theory of Games and Economic Behavior*, Princeton: Princeton University Press.

Introductory Note to Chapters 6 and 7

In his celebrated essay *Is the Scientific Paper a Fraud?* Sir Peter Meda-war argues that what finally appears in print is at best a cosmetic reconstruction of the strange ways in which the experiments came to be done. At worst it is . . . well, a fraud. Particularly in reporting motivation we tell the truth to our friends, or even to the press, but not to the scientific literature. Certainly the reader would never guess the real motivation of the work described in Chapter 6. The clue is hidden elsewhere in an article in *Computer Weekly* in which I wrote:

> The recent Oxford meeting on computer chess, as reported by Alex Bell, gave unexpected evidence that the concepts involved even in this elementary ending are not so easy to grasp. He showed six chess players a diagram of an infinitely large board with one corner (King + Rook v. King) and asked each in turn whether White can force mate. When they all said 'No' he proclaimed the superiority of human intuition over current chess programs which would be unable to make any comment on such a question. Actually the answer is 'Yes,' as pointed out some years ago by the end-game expert J. Bán (cited by T. Nemes in *Cybernetic Machines*, Iliffe Books 1969).
>
> I prescribe as Alex's penance that he should now write a program to play *optimal* King + Rook v. King, using no more than 20K bytes of store and no more than 1 second for move retrieval.

Mr. Bell is no mean programmer and he is a substantial expert on the subject of computer chess. So I felt that it behoved me to provide proof of J. Bán's assertion, and at the same time of my own capacity to perform the prescribed programming task. What follows in Chapter 6 is the result.

Chapter 7, *Chess with Computers*, develops the general theme of computer chess as a scientific study.

CHAPTER 6

King and Rook against King: Historical Background and a Problem on the Infinite Board

Mechanisations of this elementary end-game (KRK) have been done by Torres y Quevedo in the late 19th century, by B. Huberman (1968), by C. Zuidema (1974) and by M. Bramer (1975). These implementations are reviewed in the light of a distinction between 'procedural' and 'structural' approaches to embedding domain-specific knowledge.

A structural approach is described, which differs from previous KRK exercises in that it aims at a minimal-path, as opposed to a merely 'reasonable', strategy. In this paper results are reported for a problem on the infinite board.

Introduction

Machine Intelligence is concerned with programming computers for complex problem-domains, i.e. domains for which solution algorithms are known, yet are doomed by the combinatorial explosion to lose on the clock. Computer chess is a classic example. A well-known approach, first proposed in 1912 by Zermelo, is based on the idea of looking ahead along all possible paths to the end of the game. Unaided, this algorithm lacks practical applicability. Discussing it in his celebrated essay of 1950, Claude Shannon pointed out that 'a machine operating at the rate of one variation per micro-microsecond would require over 10^{90} years to calculate its first move!'

Note the qualification 'unaided'. The development of a system of artificial aids whereby domain-specific knowledge — theorems, facts, tricks, statistics, goals, evaluations and the like — can be brought to the aid of such algorithms is the defining pre-occupation of the discipline called

AI by its practitioners. But we must not lose sight here of what is general and what is specific, for the situation is quite intricate. AI seeks *general* methods of bringing *domain-specific* aids to the assistance of *general* algorithms. Thus the search algorithm to which Shannon was referring is a rather general one, of equal validity for Chess, Checkers, Go, Go-moku, Kalah, Tic-tac-toe and so forth, i.e. for any two-person zero-sum game of perfect information without chance moves. But the knowledge that permits a Grandmaster to obtain, almost without error, the same results as Shannon's 10^{90}-year computation is *specific to chess*. Indeed, the part of such knowledge that has already been codified occupies a considerable volume of print in the chess libraries of the world. According to the position maintained here, AI is the science of discovering *general* mechanisms whereby *specialised* knowledge can be made to steer *general* search algorithms, so that (against all the odds) they perform efficiently in specialised domains.

Forms of Knowledge

A graduation of forms can be recognised:

1) *Rote-memory of specific instances.* Samuel's (1959) checkers-playing program used a 'dictionary tape' to store previously encountered board positions with their computed 'back-up' evaluation scores. Since the computation of scores involved large lookahead trees great savings were attained, together with an increase in the effective depth of lookahead. Chess has a space of some 10^{45} legal positions as contrasted with about 10^{15} for checkers, so that the benefits to be obtained from rote-knowledge are in chess limited mainly to the storage of 'book openings'.

2) *'Type' positions or patterns.* An example from Zuidema is reproduced in Figure 6.1. Pattern-knowledge constitutes a greater part of the chess Master's skill — overwhelmingly so in lightning chess, which does not allow time to pursue long trains of explicit reasoning.

3) *Theorems.* An elementary example of the use of this form of knowledge is provided by the theorem that says: if in a given position a knight is the only piece free to move, he can prevent the enemy king from alternating between two opposite-coloured squares if and only if his square is opposite-coloured to that of the king. This follows from the fact that

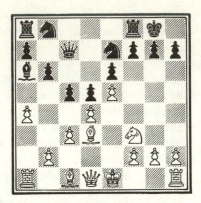

FIGURE 6.1.   Two instances of a pattern that suggests a standard mating sequence starting with a bishop sacrifice: 1 B x PCh K x B 2 N–N5ch K–N1 3 Q–R5, threatening mate on R7. In the left-hand instance the suggestion is sound, in the right-hand unsound. Zuidema (1974), from whom the example is taken, points out that pattern-knowledge has in general to be checked on each occasion by detailed lookahead.

the knight's move always changes the colours of the square on which he stands. An illustration is given in Figure 6.2, adapted from a study published in 1860 by Sam Lloyd. Black threatens to queen after N–N7ch. So White plays 1 B–R1. Because of the above theorem Black must not now capture the bishop, for after 1 . . . K x B, White's king bottles up the enemy king by 2 K–B2 (if he goes to B1, opposite in colour to the black knight's square, then we will have 2 . . . N–B4, 3 K–B2 N–N6, and Black takes the stopper out of the bottle). So instead of capturing, Black plays 1 . . . N–N7 ch, 2 K–Q2 K x B. Using the same theorem, White places his king on B1, a square of the same colour as the enemy knight. Wherever the knight now roams, he will always return with the wrong parity, and White has the black king permanently shut in.

In this paper we shall not concern ourselves with this highest level of abstraction but with what can conveniently be represented in the form of *patterns*. But we should bear in mind that one way in which a pattern can get into memory is by abstraction from sample positions to which a theorem has been successfully applied. The expert eventually recognises positions of the same *type* as Figure 6.2, abstracting from numerous episodes of reasoning. But once the pattern is there, explicit reasoning

FIGURE 6.2.   White to play and draw (study no. 49 from Bán 1963). The solution
involves use of an elementary chess theorem (see text).

is required only for checking. In domains as tractable as the one to be
considered it may not be required at all.

How to Get the Knowledge in

There are two views about where domain-specific knowledge should be
put. One says 'in data-structures', and the other says 'in procedures'.

According to the first approach we retain the solution algorithm more
or less in its pure form, and have it communicate at arm's length with a
domain-specific knowledge-base from which relevance-matching routines
digout 'advice' for the algorithm at appropriate moments. If on the other
hand we follow the procedural approach we rebuild entirely the original
naive algorithm, intimately 'embedding' domain-specific knowledge into
the constituent procedures of the rebuilt program. Devotees of the first
approach say that the second gives you a fast-running but inflexible
program, hard to read, understand, document or revise, and almost impossible to endow with powers of self-modification. Devotees of the second
approach say that the first will only become worth discussing when 'advice
programming' has been successfully applied to at least one non-trivial
problem.

There is of course no compulsion to make a cult of either the one or the other approach. Blends are possible, yet the distinction remains important. The breeding of mules, which have their own uses independent of the arts of horsemanship and the needs of donkey-work, does not nullify the labour of taxonomists. This paper re-examines a well-worn problem in chess programming with an eye to the distinction between the two approaches. After a review of earlier mechanisation of the ending king and rook against king (KRK), a problem of J. Bán for the infinite board is solved. The solution is used as an introduction to the 'structural' approach to knowledge-representation.

Earlier Mechanisations of KRK

Torres Machine

In the last decade of the 19th century Torres y Quevedo designed and built an electro-mechanical device to play the KRK game. The machine had to achieve checkmate from any starting position subject to a qualification to be explained later on. KRK is one of the easiest of the standard mates, yet it is not trivial to define an adequate strategy let alone implement it in electrical circuitry. So Torres' achievement must be ranked as a *tour de force*. The machine must also have been exceptionally durable, for Pierre de Latil reproduces a photograph of Torres' son demonstrating its operation to Norbert Wiener at the 1951 Cybernetic Congress in Paris.

Torres did not publish, but the general plan of the machine has been described by Vigneron (1914). By making common-sense interpretations in places where his text is unspecific or obscure it is possible to arrive at a reconstruction. This is shown in Table 6.1 in the form of six decision rules, checked by program as described in the Appendix. The reconstruction implies a limitation not mentioned by Vigneron, namely the initial state of the game requires to be set up so that the white rook already stands in one of the two 'zones' (see Table 6.1) and 'divides' the two kings with rank (BK) < rank (WR) < rank (WK). I have learned from Mr Alex Bell that a Torres-based machine that he saw operating in Madrid in 1958 conformed essentially to this precondition. Moreover, experiment with the strategy shows that for the worst-case starting position, shown in Figure 6.3, checkmate requires 62 moves. Nemes (1969) cites the same figure. The cumulative weight of these separate items of confirmation leaves little doubt that Table 6.1 represents a correct reconstruction.

TABLE 6.1

Reconstruction, in the form of six decision rules, of Torres' KRK machine. Ranks are assumed to be numbered downwards from the top of the board, where checkmate is finally delivered.

Precondition: divides (WR, WK, BK) and	Rules					
(file (WR) ≤ 3 or file (WR) ≥ 6)	1	2	3	4	5	6
CONDITIONS						
1. samezone (BK, WR)	Y	N	N	N	N	N
2. rank (WR) − rank (BK) > 1	–	Y	N	N	N	N
3. rank (WR) − rank (BK) = 1	–	–	Y	Y	Y	Y
4. rank (WK) − rank (BK) > 2	–	–	Y	N	N	N
5. rank (WK) − rank (BK) = 2	–	–	–	Y	Y	Y
6. file (BK) − file (WK) = 0	–	–	–	Y	N	N
7. file (BK) − file (WK) is *odd*	–	–	–	–	Y	N
8. file (BK) − file (WK) is *even* and *non-zero*	–	–	–	–	–	Y
ACTIONS						
1. WR flees horizontally from BK	X	–	–	–	–	–
2. WR advances vertically one rank	–	X	–	X	–	–
3. WK advances vertically one rank	–	–	X	–	–	–
4. WR moves horizontally one file	–	–	–	–	X	–
5. WK moves horizontally towards BK	–	–	–	–	–	X

Comments

Condition 1: 'samezone' is defined in terms of two 'zones' consisting of the files 1-3 and the files 6-8 respectively, i.e. file(BK) ≤ 3 *and* file(WR) ≤ 3 *or* file(BK) ≥ 6 *and* file(WR) ≥ 6 (leaving holes in the strategy from some starting positions – easily filled if we were to depart from Vigneron).

Condition 6: the 'direct opposition': a check from WR can force BK to yield a rank.

Condition 8: the 'distant opposition': the aim now is to convert it by a series of king moves into the direct opposition (or have the BK yield a rank before he is forced to).

Action 1: 'flees horizontally' means 'moves so as to maximise the file distance between BK and WR'. Vigneron is not specific.

Action 2: as part of Rule 2 this effects a step-by-step advance until WR 'confines' BK (i.e. until Condition 3 becomes rule). In Rule 4 the same Action gives check in response to Condition 6.

Action 3: 'vertically' is an assumption. Vigneron is not specific.

Action 4: a 'waiting move' to change the parity of the inter-king file distance. When a choice exists, prefer to move away from BK.

The strategy shown introduces some of the main concepts of the KRK game, and demonstrates how few are needed if the aim is merely to arrive at checkmate, irrespective of the length of the path. Fine's (1941) conjecture that with White to move mate can be guaranteed in at most seventeen moves (33 ply) has been replaced by the definitive result 31 ply, obtained by exhaustive enumeration (Clarke, 1977). Compare with Torres' 123-plus (the 'plus' standing for additional moves concerned with establishing the precondition). The multiplication and refinement of concepts for progression from a Torres-like strategy to an optimal-path strategy (defined under the minimax condition that White always plays to hasten the end and Black to delay it) is a matter of some interest.

In Table 6.1 Torres' eight conditions are separated by horizontal lines into four groups corresponding to four broad strategic concerns.

First group: is the rook actually or potentially in danger? If so, then the rook should flee, says rule 1.

Second group: does the rook 'confine' the black king, i.e. bar him from retreating one rank? If not, keep advancing the rook, says rule 2.

Third group: is the inter-king rank distance small enough to begin manoeuvring for the 'opposition'? If not, then keep advancing the white king, says rule 3.

Fourth group: does the white king possess the direct opposition? If so, then exploit it with check, says rule 4. If not, then perhaps the distant opposition? Then try for the direct opposition, says rule 6. Otherwise rule 5 prescribes a waiting move.

The simplicity of Torres' decision logic underlies both (a) the extraordinary compactness with which the algorithm can be represented (logical redundancy among the conditions allows us to delete three rows from the upper part of Table 6.1 if we wish) and (b) the extraordinary inelegance of its play. Subsequent mechanisers have achieved gains under heading (b) at the expense of losses under heading (a). One of them, Coen Zuidema, an International Chess Master and computer scientist, remarks: 'A small improvement entails a great deal of effort.' We now turn to his study of KRK.

Zuidema's Program

It is a normal approach to any tricky problem, in AI as elsewhere, to start with a crude, rough-hewn, Torres-like strategy and then to improve it piece-meal by repeated cycles of testing and patching, as in the

1	R–K2	(5)	K–B6		12	R–N3ch	(4)	K–N5
2	K–N1	(6)	K–Q6		13	K–N2	(3)	K–B5
3	R–KR2	(1)	K–B6		14	R–QR3	(5)	K–K5
4	R–KN2	(5)	K–Q6		15	K–B2	(6)	K–Q5
5	K–B1	(6)	K–K6		16	K–K2	(6)	K–B5
6	K–Q1	(6)	K–B6		17	R–KR3	(1)	K–Q5
7	R–QR2	(1)	K–K6		18	R–KN3	(5)	K–B5
8	R–QN2	(5)	K–B6		19	K–K2	(6)	K–N5
9	K–K1	(6)	K–N6		20	K–B2	(6)	K–R5
10	K–B1	(6)	K–R6		21	K–N2	(6)	K–N5
11	K–N1	(6)	K–N6		22	R–N4ch	(4)	K–N4

etc.

The sequence-pattern exhibited by moves 13–22 is repeated four more times; adding the first twelve moves, we have 5 x 10 + 12 = 62.

FIGURE 6.3.   An adverse initial position for the Torres strategy, requiring 62 moves by White to force checkmate. The move sequence is given underneath in the diagrammed board position; numbers in parentheses identify rules from Table 6.1.

debugging of computer programs (see Sussman 1974). But note in passing the debugging is ordinarily directed towards a correct implementation of a *given* strategy rather than towards the strategy itself.

Optimising a strategy by trial and error hinges critically on the detection of exceptions. It would be tempting for example to introduce into Table 6.1 a refinement of rule 1 to detect in the initial position the protection of the rook by the white king as a substitute for fleeing.

Zuidema's (1974) study describes two ALGOL 60 programs for KRK. The second represents an elaboration designed to correct some of the inelegancies of play exhibited by the first. As worst-case behaviour the primitive version required twenty-seven moves (53 ply) to mate from one particular position (Zuidema's figure 29) for which the length of the minimax path is known from Clarke's work to be fourteen moves (27 ply).†

The improvement of play in the second version (the ALGOL text of which is given in full) is not easy to assess from Zuidema's account. For him, however, it was evidently purchased at too high a price: 'The conclusion forces itself that refining the algorithm and exceptions of rules give rise to an overburdened program . . .' and in his Introduction: 'A more refined strategy is needed to elevate level of play. A small improvement, however, entails a great deal of expense in programming effort and program length. The new rules will have their exceptions too.'

Table 6.2 gives the total bulk, in 60-bit words, of compiled code for Zuidema's old and new versions. The procedure 'room' described below contains nearly all the domain-specific knowledge.

Torres' strategy does not look ahead, but selects its moves directly on the basis of the current position. By contrast, Zuidema's does not prune the number of possible moves down to only one, but applies some milder preliminary criteria to produce a candidate set. Pruning is at three levels:

1) *Unconditional.* Example: rook moves to ranks 2, 3 and 4 and to files 2 and 3 are excluded: the program transposes every position so that the black king stays in the triangle e5–e8–h8 (algebraic notation). The move to rank 1 is not excluded so that it can be used as a 'waiting move' when required. The fact that waiting moves *are* required in KRK should be noted. If Black is allowed null moves then White cannot force the win.

2) *Conditional on features of the current position.* Example: if the inter-king rank distance is more than three, then a king move is always played unless the rook is *en prise*.

3) *Conditional on features of the successor position.* Example: a king move which would enlarge the inter-king rank distance is excluded.

The candidate moves that pass this three-level filter are then applied to generate a set of successor positions, and these are scored by a purely

---

†White's first error was his third move, RG4. Kd1 was necessary.

TABLE 6.2

Program length, and some of its components, in C. Zuidema's old and new versions (units are 60-bit words of object code).

	old	new
total program	2000	2900
input-output	715	715
the procedure 'room'	280	1070

static evaluation function (i.e. no further lookahead is employed in calculating it). This function is computed by the procedure 'room' based, with many elaborations and refinements, on counting the number of squares over which the black king would be free to roam if the white pieces sat still. The concept is an advance on those of Torres and leads to more purposeful 'hemming in' of the enemy. The 'room' idea is also of general chess utility beyond the bounds of KRK.

Huberman's Program

In 1968 Barbara Huberman wrote a program to play the endings king and rook against king, king and two bishops against king, and king, bishop and knight against king.

Confining attention to the first of these, play was respectable, although not quite as economical as Zuidema's new version. The principles on which it was constructed were totally different, and are of interest in exemplifying in pure form what we have termed the 'structural' approach. Huberman's base of domain-specific knowledge was a set of logic formulae expressing 'goals'. The domain-independent solution algorithm took the form of a procedure for constructing a 'forcing tree' from which playing sequences satisfying the goals could be generated. Huberman's study provided a convincing demonstration of the versatility achievable by the structural approach: the same program with a different assertion-base coped successfully with the formidable ending K + B + N against K.

Bramer's Program

Recently M.A. Bramer (1975) has reported an experiment with the KRK endgame. His approach has features in common with Huberman's but the salient differences are: (a) Bramer's definition and ordering of goal patterns (he calls them 'equivalence classes') is such that no search at all is required, the principle of play being simply 'in any position, choose the move which gives the most favourable successor position'. (b) The quality of play is much better. Bramer states that 'no case has so far been found where the program takes more than the theoretical maximum necessary number of 17 moves before checkmate . . .'. Since we know from Clarke (this volume) that the theoretical maximum is in fact sixteen, we can infer that Bramer's program generates sub-optimal play, while accepting his assurance that it is at least near-optimal.

Current Work

Optimal KRK

In the restricted KRK context Huberman's work represents the use of a steam-hammer to crack a nut. Accordingly we consider the question: can we design a nut-sized hammer on similar principles and use it to do what Huberman did not do (because she did not attempt it), namely to build an *optimal* KRK strategy in the minimax sense defined earlier. We shall require the program modules and data elements to be few and simple. We shall also require the process of building and re-building the strategy ('programming effort' in Zuidema's terms) to be straightforward and not burdensome.

In a step-wise approach to the final program, a version was developed for the problem of mating the lone king on an infinite board with one corner.

The Infinite Board

On a board which extends without limit in all directions the black king cannot be mated even if he co-operates. This follows from the fact that in a checkmate position every square of the 3 x 3 array shown in Figure 8.4 must be attacked by white pieces. The white king can attack at most three squares adjacent to the enemy king and the white rook can attack at most three more. We are left with a deficit of three squares.

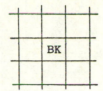

FIGURE 6.4. The black king cannot be in checkmate from WK and WR on an infinite edge-less board. All nine squares shown above must be controlled, exceeding the combined powers of the two attacking pieces.

On an infinite board with a single edge (or with two parallel edges) the mate can only be forced from a few special starting positions: in the general case the black king can be brought to checkmate only if he co-operates. The presence of at least one corner is thus a necessary condition for mate to be forceable regardless of starting position. The reasons are not immediately obvious. An outline proof follows which introduces notations and methods basic to the rest of this study. We later show the condition to be sufficient as well as necessary.

The left-hand diagram of Figure 6.5 shows the sole checkmate pattern possible on the infinite board with an edge but no corners. The uniqueness of the checkmate pattern follows from the fact that the edge abolishes three of the nine squares of Figure 6.4 and the WK and WR each deploys his maximum competence in taking care of three each of the six remaining. No other arrangement of pieces allows this.

Proof that legal starting positions exist from which White cannot force checkmate in any sequence of steps proceeds by noting that all forcing sequences terminate in pattern 0 (left-most in Figure 6.5). Hence they can all be systematically generated by building a derivation tree backwards from pattern 0, first by constructing all patterns from which a white-optimal move leads in a single step to pattern 0, then by constructing all the patterns from which a black-optimal move leads in a single step to one or another of the level-1 patterns already generated, and so on. It turns out that the tree-construction soon terminates in parentless nodes after only ten new patterns have been added to pattern 0. This is shown in Figure 6.5, which also contains a commentary on the various arcs of the tree. These mark pattern-to-pattern transitions. The eleven patterns do not exhaust the set of patterns required to describe all legal positions of the

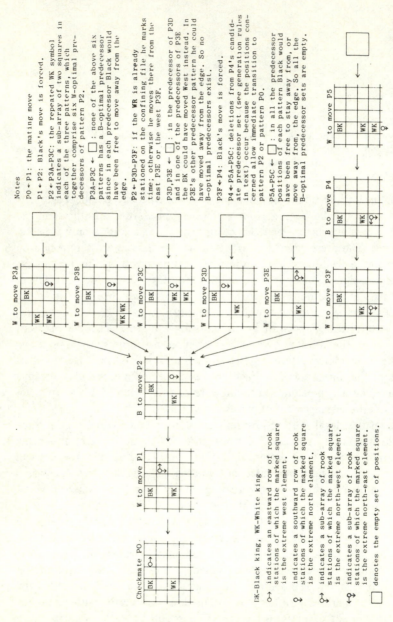

Notes

P0 ← P1: the mating move

P1 ← P2: Black's move is forced.

P2 ← P3A–P3C: the repeated WK symbol indicates a sub-array of two squares in each of the three patterns, which together comprise six W-optimal predecessors of pattern P2.

P3A–P3C ← ☐ : none of the above six patterns has a B-optimal predecessor since in each pattern Black would have been free to move away from the edge.

P2 ← P3D–P3F: if the WR is already stationed on the confining file he marks time; otherwise he moves there from the east P3E or the west P3F.

P3D,P3E ← ☐ : in the predecessor of P3D and in one of the predecessors of P3E the BK could have moved West instead. In P3E's other predecessor pattern he could have moved away from the edge. So no B-optimal predecessors exist.

P3F ← P4: Black's move is forced.

P4 ← P5A–P5C: deletions from P4's candidate predecessor set (see generation rules in text) occur because the positions concerned allow immediate transition to pattern P2 or pattern P0.

P5A–P5C ← ☐ : in all the predecessor positions of each pattern Black would have been free to stay away from, or move away from, the edge. So all the B-optimal predecessor sets are empty.

BK—Black king, WK—White king

↻→   indicates an eastward row of rook stations of which the marked square is the extreme west element.

↻   indicates a southward row of rook stations of which the marked square is the extreme north element.

↻→   indicates a sub-array of rook stations of which the marked square is the extreme north-west element.

↻→   indicates a sub-array of rook stations of which the marked square is the extreme north-east element.

☐   denotes the empty set of positions.

FIGURE 6.5.   Tree of optimal derivations of the checkmate pattern for infinite cornerless KRK. The board is bounded to the north by a single edge.

infinite cornerless **KRK** problem. They do not, for example, include cases where the BK is not on the edge, nor cases where the inter-king file distance exceeds two, nor cases where the inter-king rank distance exceeds three. There exist, therefore, legal positions not represented on the tree, and from such positions no forcing sequences to checkmate can be found.

In order to convert the annotated figure into a satisfactory demonstration, it remains to specify rules for generating W-optimal and B-optimal predecessor patterns from even-numbered and odd-numbered patterns respectively. A W-optimal predecessor of a pattern is one which is linked to that pattern by a white-optimal move; a B-optimal predecessor of a pattern is one which is linked by a black-optimal move. Strict definitions are embodied in the following generating rules.

Even-numbered patterns (Black to move):

1) For each member of the given pattern p, construct the set of legal positions from which that member can be generated by a white move.

2) Form the set-union of all the sets thus computed. Call this the candidate set.

3) For each member of the candidate set generate its legal successors. If any of these matches with some pattern already on the tree and closer to the root than p, then delete that member from the candidate set.

4) Call the residual set the set of W-optimal predecessors of p, and represent it by one or more patterns.

Odd-numbered patterns (White to move):

1) For each member of the given pattern p', construct the set of legal positions from which that member can be generated by a black move.

2) Form the set-union of all the sets thus computed. Call this the candidate set.

3) For each member of the candidate set generate its legal successors. If any of these fails to match with p', then delete that member from the candidate set.

4) Call the residual set the set of B-optimal predecessors of p', and represent it by one or more patterns.

The rules stated require extension since they call for an operation to be performed on each member of a pattern, which (since the range of rook-moves is unbounded) may have infinitely many members. The generating algorithm must therefore be able to classify members of a pattern as the same if they are distinguished only by different rook-stations in a manner which is irrelevant to play. A sufficient classification rule is to regard as the same for purposes of pattern-description all rook-stations that lie on the same rank or file at a distance of three or more squares from the nearest other piece.

Positions, Sets and Patterns

A position is represented as a list of three co-ordinate pairs, for the BK, the WK and the WR respectively. For the BK the x co-ordinate has the constant value *undefined* (written u or UND) on the infinite cornerless board, while the y co-ordinate has the value 0 if the BK is on the edge, and $1, 2, 3, \ldots$ if it is $1, 2, 3, \ldots$ squares from the edge. The WK and WR have their co-ordinates reckoned *from the square of the BK*, so that the pair (i,j) denotes a square lying i steps to the east and j steps to the south of the BK.

A pattern selects three sub-arrays of the board, a BK-array, a WK-array and a WR-array, each of which specifies a set of possible stations for the relevant piece. The BK-array and WK-array are typically, but not necessarily, unit arrays, i.e. single squares. On the infinite board the WR-array can be of infinite extent, but the infinite set of positions is describable by a finite set of patterns. The representation of a pattern is similar to that of a position except that in place of a pair of co-ordinates for each piece, specifying its station, we have a pair of intervals, specifying the sub-array within which its station lies. Thus we have the following representations:

> position 0 $(((u,u),(0,0)), ((0,0),(2,2)), ((2, \infty),(0,0)))$
> position 1 $(((u,u),(0,0)), ((0,0),(2,2)), ((2, \infty),(1, \infty)))$
> position 2 $(((u,u),(0,0)), ((-1, -1),(2,2)), ((1, 1),(2, \infty)))$.

To detect whether a position is a member of a given pattern we notionally generate the Cartesian product of the latter's three sub-arrays and check whether any of the triples so generated is identical with the triple of co-ordinate pairs which specifies the position. As a practical computation it is only necessary to check that each co-ordinate in the position-specification falls within the corresponding interval of the pattern. For proceeding in the inverse direction we require a rule that will generate from a set of positions a minimal defining set of patterns. To do this without ambiguity is not possible except by introducing arbitrary criteria. For example, there are four ways of partitioning the six WK-stations of patterns P3A–P3C into three sub-arrays of which the way followed in Figure 6.5 is only one. The criterion adopted was to make the partitions as near equal in size as possible by maximizing the product of their sizes.

Optimal Infinite Cornerless KRK

Figure 6.5 not only outlines a proof: for the sparse set of positions from which mate can be forced it specifies an optimal strategy for both players. Let us consider a way of representing the knowledge expressed by the figure in a form convenient for a program to handle.

In addition to a base of patterns, we construct a table of *advice triples* each of the form:

POINTER TO CONDITION PATTERN	PATTERN SPECIFYING PLAUSIBILITY CLASS	POINTER TO GOAL PATTERN(S)

To retrieve a correct move, the input position is matched against the pattern addressed by the left-hand element of each stored advice triple until a condition-match is found. Legal moves permitted by the 'plausibility class' are then applied to the position in turn, matching the result in each case against the pattern addressed by the triple's right-hand member. As soon as a goal-match is found, the corresponding move is output. The complete set of patterns and advice is set out in Table 6.5. To run this as a strategy on the machine we require procedures for generating legal moves, for matching positions with patterns, and for input and output. POP-2 functions were de-bugged on this trivial example, and then used for developing a base of patterns and advice for the infinite board with one corner.

TABLE 6.3

Pattern-base and advise list for optimal KRK on the infinite cornerless board — see text and Figure 6.5. Meanings of symbols are as in Table 6.5, 6.6. The last symbol in each row of the advice list gives the number of steps to mate.

```
P0   UND  UND   0   0    0    0    2    2  MIF  -2    0    0
P1   UND  UND   0   0    0    0    2    2  MIF  -2    1  INF
P2   UND  UND   0   0   -1   -1    2    2    1    1    2  INF
P3A  UND  UND   0   0   -2   -2    1    2    1    1    2  INF
P3B  UND  UND   0   0   -2   -1    3    3    1    1    2  INF
P3C  UND  UND   0   0    0    0    2    3   -1   -1    2  INF
P3D  UND  UND   0   0   -1   -1    2    2    1    1    1  INF
P3E  UND  UND   0   0   -1   -1    2    2    2  INF    2  INF
P3F  UND  UND   0   0   -1   -1    2    2  MIF   -1    3  INF
P4   UND  UND   0   0    0    0    2    2  MIF    0    3  INF
P5   UND  UND   0   0    0    0    2    3    0    0    3  INF


0  P0   0  0  0  0  0  0  0  0  0  0  0  0   []       0
1  P1   0  0  0  0  0  0  0  0  1  0  0  0   [P0]     1
0  P2   0  0  0  1  0  0  0  0  0  0  0  0   [P1]     2
1  P3A  0  0  1  0  0  1  0  0  0  0  0  0   [P2]     3
1  P3B  1  0  0  0  1  0  0  0  0  0  0  0   [P2]     3
1  P3C  0  0  0  1  0  0  1  0  0  0  0  0   [P2]     3
1  P3D  0  0  0  0  0  0  0  0  1  1  0  0   [P2]     3
1  P3E  0  0  0  0  0  0  0  0  0  0  0  1   [P2]     3
1  P3F  0  0  0  0  0  0  0  0  0  1  0  0   [P2]     3
0  P4   0  0  1  0  0  0  0  0  0  0  0  0   [P3F]    4
1  P5   1  0  0  0  0  0  0  0  1  1  1  1   [P4]     5
```

Optimal Infinite KRK with One Corner

On the infinite board with two edges meeting at a corner mate *can* be forced. A problem originally posed by J. Bán is as follows. Suppose that the white king and rook are in the one and only corner of an infinite chessboard, while the black king is in an arbitrarily remote square; is it possible to checkmate the black king? T. Nemes (1969), from whose account of the foregoing formulation is taken, continues: 'The analysis of this problem requires consideration of a topologico-geometrical nature. To mate the black king, it is necessary to drive it to one edge of the board; consequently, the white rook and king must be on the 'far' side of the black king. However, a slightly more profound analysis shows that, by moving the rook to the far side of the black king, we have confined the latter in a rectangle (one of whose diagonals is subtended by the rook and the finite corner of the board). Now the white king can start and leave this rectangle (it is, of course, not confined by its own rook!) and move on until it is on the far side of the black king in both the x and the y direction. Now both the white chessmen are on the far side of the black king and the black king can be driven into the finite corner and check-mated.'

A.J. Roycroft (personal communication) comments: 'When I first came across this I was unconvinced, as it seemed to say that after White has positioned his rook he need only move his king', but rightly concluded that whatever the gaps in Nemes' reasoning, he was essentially correct. Table 6.4 shows a strategy for achieving the first phase – getting the white chessmen to the far side of the black king – which is complete for the initial conditions shown in the table.† Under these conditions the strategy is also optimal according to the following informal reasoning.

The strategy covers the first phase of 'one-corner infinite KRK', and is concerned with creating a situation in which the BK has no move that can further increase his distance from the corner in 'city-block' units, i.e. the sum of his co-ordinates x_{BK}, y_{BK}. This requires that WK and the WR to have got 'beyond' the BK and the WR to be safe from capture. The condition is satisfied by the situation on achievement of 'goalpattern 5' of Table 8.4, an example of which can be seen in Figure 6.6 where the results of running an implementation of the strategy are reproduced. We first show that it is not possible to achieve such a situation in a smaller number of moves in the case that Black follows a consistent policy of flight. We then show that departures from that policy cannot gain Black anything.

The fundamental principle is that the BK can be overtaken by the WK if and only if he can be constrained to flee along lines of latitude and longitude while his pursuer is free to move diagonally. At the expense of the three rook moves White establishes just this constraint, so that if Black makes only flight moves each one of them has to be *either* due east or due south (i.e. either x_{BK} or y_{BK} is incremented, but not both). White further ensures, so long as each black move is a flight move (i.e. one that increases $x_{BK} + y_{BK}$), that all his own king moves are diagonal to the south-east. Each such move increments *both* x_{WK} *and* y_{WK}, so that when it is followed by a black flight move a net increment of one move is gained. This represents the maximum rate at which the fleeing BK can be overtaken. Remembering that Black steals extra flight moves following each of White's three rook moves, we can reckon from the initial positions

---

† The condition omitted is that in which the BK stands on, or immediately adjacent to, the board's single diagonal. It is possible, although tedious, to show that a suitable complication of the same strategy will meet this case.

```
      Ø   1   2   3   4

Ø   WK   .   .   .   .

1    .  WR   .   .   .

2    .   .   .   .  BK
```

WHITE TO PLAY


```
BANØ
BAN1 ——— WR sets up an east-west line in the south
BAN1
BAN2 ——— WR goes east
BAN24
BAN2 ——— WK starts trek to south-east
BAN24
BAN24
BAN2
BAN24
BAN2
BAN24
BAN2
BAN3 ——— WR moves south
BAN4  ⎫
BAN5  ⎪
BAN4  ⎪
BAN5  ⎪
BAN4  ⎬ Second leg of WK's trek to south-east
BAN5  ⎪
BAN4  ⎪
BAN5  ⎪
BAN4  ⎪
BAN5  ⎭
BAN4 ——— BK backs off north, giving the following position:
```


```
     Ø   1   2   3   4   5   6   7   8   9  1Ø  11

Ø    .   .   .   .   .   .   .   .   .   .   .   .

1    .   .   .   .   .   .   .   .   .   .   .   .

2    .   .   .   .   .   .   .   .   .   .   .   .

3    .   .   .   .   .   .   .   .   .   .   .   .

4    .   .   .   .   .   .   .   .   .   .   .   .

5    .   .   .   .   .   .   .   .   .   .   .   .

6    .   .   .   .   .   .   .   .   .   .  BK   .

7    .   .   .   .   .   .   .   .   .   .   .   .

8    .   .   .   .   .   .   .   .   .   .   .   .

9    .   .   .   .   .   .   .   .  WK   .  WR
```

Matches with
Pattern PW8, marking
the start of part B.
BK is fenced in,
WK is safely on
the fence.

WHITE TO PLAY

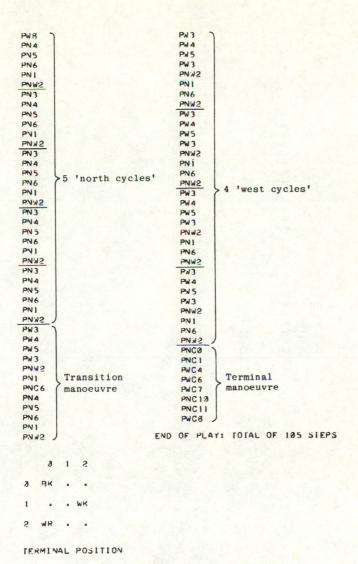

FIGURE 6.6(a).   Output of the program when applied to an instance of Bán's problem (top left). Print-out shows successive pattern matches, with annotations corresponding to the categories of Tables 6.4 and 6.7.

```
        0  1  2  3  4  5

0   WK  .  .  .  .  .

1    .  WR  .  .  .  .

2    .  .  .  .  .  BK
```

WHITE TO PLAY


```
BAN0
BAN1
BAN1
BAN2
BAN2A
BAN2
BAN2A
BAN2
BAN2A
BAN2
BAN2A
BAN2
BAN3
BAN4
BAN5
BAN4
BAN5
BAN4
BAN5
BAN4
BAN5
BAN4
BAN5
BAN4
BAN5
BAN4
```

```
        0  1  2  3  4  5  6  7  8  9  10  11  12

0    .  .  .  .  .  .  .  .  .  .  .   .   .

1    .  .  .  .  .  .  .  .  .  .  .   .   .

2    .  .  .  .  .  .  .  .  .  .  .   .   .

3    .  .  .  .  .  .  .  .  .  .  .   .   .

4    .  .  .  .  .  .  .  .  .  .  .   .   .

5    .  .  .  .  .  .  .  .  .  .  .   .   .

6    .  .  .  .  .  .  .  .  .  .  .   .   .

7    .  .  .  .  .  .  .  .  .  .  .  BK   .

8    .  .  .  .  .  .  .  .  .  .  .   .   .

9    .  .  .  .  .  .  .  .  .  .  .   .   .

10   .  .  .  .  .  .  .  .  .  .  WK  .  WR
```

WHITE TO PLAY

```
PW8        PW3        PW3
PN4        PW4        PW4
PN5        PW5        PW5
PN6        PW3        PW3
PN1        PNW2       PNW2
PNW2       PN1        PN1
PN3        PNC6       PN6
PN4        PN4        PNW2
PN5        PN5        PW3
PN6        PN6        PW4
PN1        PN1        PW6
PNW2       PNW2       PNC0
PN3        PW3        PNW2
PN4        PW4        PN1
PN5        PW5        PNC11
PN6        PW3        PWC8
PN1        PNW2       PNC12
PNW2       PN1        PNC11
PN3        PN6        PNC10
PN4        PNW2       PNC10
PN5        PW3        PNC11
PN6        PW4        PWC8
PN1        PW5
PNW2       PW3        END OF PLAY:
PN3        PNW2
PN4        PN1        TOTAL OF 119 STEPS
PN5        PN6
PN6        PNW2
PN1        PW3
PNW2       PW4
PN3        PW5
PN4        PW3
PN5        PNW2
PN6        PN1
PN1        PN6
PNW2       PNW2
```

```
         0   1   2

0   BK   .   .

1        .   .   WK

2   WR   .   .

TERMINAL POSITION
```

FIGURE 6.6(b).   Output obtained from a starting position differing in parity to that of Figure 6.7a. This difference is the cause of the divergence of the two terminal sequences, shown in detail in Figures 6.12a and 6.12b.

TABLE 6.4

Demonstration, in the form of a complete strategy, that on the infinite board with one corner the pre-conditions can be created for the standard mating sequence (see text).

	Patterns	Moves for White	Notes

initial:	file(WK) = rank(WK) = 0;
	 file(WR) = rank(WR) = 1;
	 filediff(BK, WR) ≥ rankdiff(BK, WR) + 2
		southward rook-move achieving goal-pattern 1.	1

goal-pattern 1:	rankdiff(WR, BK) = 1	eastward rook-move achieving goal-pattern 2.	2
goal-pattern 2:	filediff(WR, BK) = 3 + rankdiff(WR, WK)	if possible then south-east king-move otherwise southward king-move; repeat until goal-pattern 2 is achieved.	3
goal-pattern 3:	rankdiff(WK, WR) = 1	if filediff(WR, BK) > 2 then westward rook-move achieving filediff(WR, BK) = 2 otherwise southward rook-move achieving goal-pattern 4.	4
goal-pattern 4:	rankdiff(WR, BK) = filediff(WR, WK)	if goal-pattern 5 then exit otherwise south-east king-move; repeat until goal-pattern 5 is achieved.	5
goal-pattern 5:	rankdiff(WK, WR) = 0	exit	

Notes:
1. The rook sets up an east-west line in the south.
2. The rook goes east to be out of the black king's reach during the white king's first trek.
3. The white king treks south and east to cross the rook's line.
4. The rook goes further south to be out of the black king's reach during the white king's second trek.
5. The white king treks south-east to rejoin the rook; this time the black king is completely fenced in and the white king is safely on the fence.

of the two kings the minimum number of net increments needed to create the pre-condition, and this is equal to the total number of moves to be made by the WK. The required expression is m_{WK} = filediff(BK,WK) + rankdiff(BK, WK) + 3.

Before accepting m_{WK} as White's path-length under an optimal strategy we now ask whether the BK must necessarily flee at each step. Might he not do better at some stage to switch from increasing $x_{BK} + y_{BK}$ at each move, to going back and harassing the movements of the WK? The reasons why this does not work are as follows. The only achievable harassment consists in obliging the WK to move south rather than south-east during his first trek. This might be thought to render more lengthy White's subsequent task of bringing the WK and the WR together. But the loss is more than compensated by the westward rook move after goal-pattern 3 has been achieved. Hence the length of the southward trek (and the eventual value of y_{BK}) is not increased by the BK's manoeuvres. The eventual value of x_{BK}, on the other hand, has been diminished by twice the number of westward moves made by the BK. Thus, if the BK makes moves which are other than flight moves, he not only fails to prolong phase 1 but also stores up for himself a penalty through entering upon phase 2 in a smaller confining rectangle than he need have done.

To complete the argument we have to show that the WR cannot impose the 'latitude-or-longitude' constraint on Black's flight moves at the expense of fewer than three moves. One move is necessary to set up the constraining barrier to the south, and the second rook move necessarily follows this barrier to the east if the constraint is to be maintained. This, however, leaves the BK separated from the WR by fewer files than the WK, so that the BK cannot be impeded from advancing along the barrier and capturing the WR. Therefore at least one more rook move must be required.

The next task is to chase the black king to the corner. This proceeds in a series of cycles. There is a cycle of six steps that forces the black king one rank to the north, illustrated in Figure 6.7. If he decides to flee due west instead of north, there is a cycle of eight steps, shown in Figure 6.8, that forces him *two* files to the west, i.e. one file per four steps.

Alternating these two cycles according to the black king's direction of flight, we ask how many steps are required to bring the black king from a square with co-ordinates x, y to the corner. The black king will so plan his flight as to postpone the moment of contact with either edge. The reason is that once he is on an edge White can set up conditions for driving

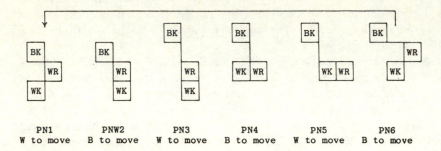

PN1   PNW2   PN3   PN4   PN5   PN6
W to move   B to move   W to move   B to move   W to move   B to move

FIGURE 6.7.   The 'north cycle' of six steps that forces Black to yield one rank
                per cycle.

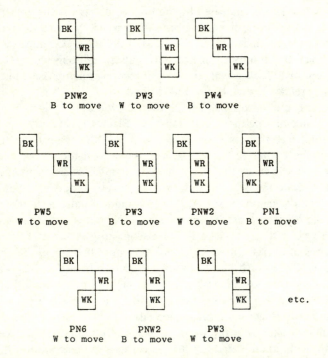

PNW2   PW3   PW4
B to move   W to move   B to move

PW5   PW3   PNW2   PN1
W to move   B to move   W to move   B to move

PN6   PNW2   PW3
W to move   B to move   W to move   etc.

FIGURE 6.8.   The 'west cycle' of eight steps which forces the black king to yield
                two files per cycle.

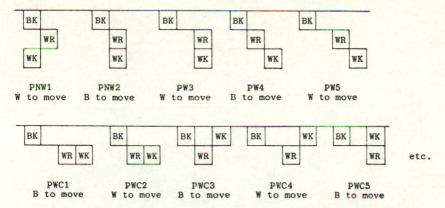

FIGURE 6.9.   If the black king were to hit the north edge in the course of a 'north cycle' (optimal KRK avoids this) the WR would set up an east-west corridor one square wide, along which the BK can be chased at the rate of one file per two steps.

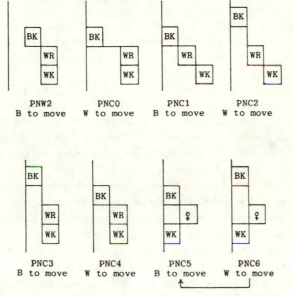

FIGURE 6.10.   When the black king emerges from a west cycle on to the west edge, as occurs above at PNC0, the WR sets up a north-south corridor one square wide along which the WK chases the BK at the rate of one rank per two steps.

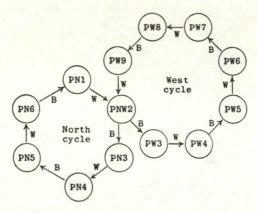

FIGURE 6.11.   Graphic representation of the two forcing cycles. The arcs are labelled according to the move, and the nodes according to the conventions of Figures 6.7 and 6.8.

him towards the corner at a much faster rate of one rank (or file) per 2 steps. This is shown in Figures 6.9 and 6.10. So if the north cycle and west cycle can be treated independently, the black king can clearly be brought to an edge in at most $4x + 6y$ steps.

Figure 6.11 shows the north and west cycles graphically. There is but one single position belonging to both cycles allowing Black to transfer, if he chooses, from one cycle to the other. It follows that however he exercises this option he must yield one rank for every six steps round the north cycle and two files for every eight steps round the west cycle. The cycles can be treated independently.

Table 6.5 shows the advice triples corresponding to the complete optimal solution of Bán's problem, operating on the pattern base set out in Table 6.6. As in the cornerless case, to execute this advice the input position is tested against the condition problems until a match is found. A move compatible with the associated plausibility class is then generated and applied, and the resulting position is tested against the triple's goal pattern. If it matches, then the move is output. If not, the next compatible move is tried. In order to accommodate the predicates of Table 6.4, the apparatus for specifying a co-ordinate interval was extended, as described in the caption of Table 6.6.

TABLE 6.5

Advice list for solution of Bán's problem. The first digit of each piece of advice,
1 or 0, denotes whether it is W or B to play. Then follows the identifier of the left-
hand, or 'condition', pattern. The sequence of twelve binary digits that follows
defines the 'plausibility class' according to the code: North South East West
NE SE NW SW North South East West. The first eight symbols refer to king moves
(W or B as the case may be). The last four symbols refer to rook-moves in the case
that W is to play; otherwise they are set to zero. The 'target list' follows, enclosed
in square brackets. The last component, 'steps to mate', is not shown.

```
1 BAN3    0 0 0 0 0 0 0 0 0 1 0 0   [BAN1]
0 BAN1    0 0 1 0 0 0 0 0 0 0 0 0   [BAN1]            ⎫
1 BAN1    0 0 0 0 0 0 0 0 0 0 1 0   [BAN2]            ⎪
0 BAN2    0 0 1 0 0 0 0 0 0 0 0 0   [BAN2A BAN3]      ⎪ part A:
1 BAN2A   0 0 0 0 0 1 0 0 0 0 0 0   [BAN2]            ⎬ outward bound
1 BAN3    0 0 0 0 0 0 0 0 1 0 0 0   [BAN4]            ⎪
0 BAN4    1 1 0 0 0 0 0 0 0 0 0 0   [BAN5 PW8]        ⎪
1 BAN5    0 0 0 0 1 1 0 0 0 0 0 0   [BAN4]            ⎭
1 PW8     0 0 0 0 1 0 0 0 0 0 0 0   [PN4]               transitional
0 PN4     0 0 0 1 0 0 0 0 0 0 0 0   [PN5]            ⎫
1 PN5     0 0 0 0 0 0 0 1 0 0 0 0   [PN6]            ⎪
0 PN6     0 0 1 0 0 0 0 0 0 0 0 0   [PN1]            ⎬ part B:
1 PN1     0 0 1 0 0 0 0 0 0 0 0 0   [PNW2]           ⎪ north cycle
0 PNW2    1 0 0 1 0 0 0 0 0 0 0 0   [PN3 PW3 PNC0]   ⎪
1 PN3     0 0 0 0 0 0 1 0 0 0 0 0   [PN4]            ⎭
1 PW3     0 0 0 0 0 0 0 0 1 0 0 0   [PW4]            ⎫
0 PW4     1 0 0 1 0 0 0 0 0 0 0 0   [PW6 PW5]        ⎪
1 PW5     1 0 0 0 0 0 0 0 0 0 0 0   [PW3 PNC0]       ⎪ part C:
0 PW3     0 1 0 0 0 0 0 0 0 0 0 0   [PNW2]           ⎬ west cycle
1 PNW2    0 0 0 1 0 0 0 0 0 0 0 0   [PN1]            ⎪
0 PN1     1 0 0 1 0 0 0 0 0 0 0 0   [PN6 PNC6 PNC11] ⎪
1 PN6     0 0 0 0 0 0 0 0 0 0 0 1   [PNW2]           ⎭
1 PNC6    1 0 0 0 0 0 0 0 0 0 0 0   [PN4]               transitional
1 PW6     0 0 0 1 0 0 0 0 0 0 0 0   [PNC0]
0 PNC0    0 0 1 0 0 0 0 0 0 0 0 0   [PNW2]           ⎫
1 PNC11   0 0 0 0 1 0 0 1 0 0 0 0   [PWC8]           ⎪
0 PWC8    1 0 0 0 0 0 0 0 0 0 0 0   [PNC12]          ⎪
1 PNC12   1 0 0 0 0 0 0 0 0 0 0 0   [PNC11]          ⎪
0 PNC11   0 0 1 0 0 0 0 0 0 0 0 0   [PNC10]          ⎬ terminal
1 PNC10   0 0 0 0 0 0 0 0 0 1 0 0   [PNC10]          ⎪ manoeuvres
0 PNC10   1 0 0 0 0 0 0 0 0 0 0 0   [PNC11]          ⎪
1 PNC0    0 0 0 0 0 0 0 0 0 0 1 0   [PNC1]           ⎪
0 PNC1    1 0 0 0 0 0 0 0 0 0 0 0   [PWC4 PNC2]      ⎪
1 PWC4    1 0 0 0 0 0 0 0 0 0 0 0   [PWC6]           ⎪
0 PWC6    0 1 0 0 0 0 0 0 0 0 0 0   [PWC7]           ⎪
1 PWC7    1 0 0 0 0 0 0 0 0 0 0 0   [PNC10]          ⎭
```

The advice shown only caters for positions which can arise in optimal
play. Output from two specimen runs of the program was given in Figure
6.6. Figure 6.12 shows the last few steps in these sequences run through
with output parameters set to give an 'introspective record', and Figure
6.13 shows the flow of control through the complete set of advice. Total
POP-2 text (exclusive of patterns and advice) was 685 lines.

TABLE 6.6

Pattern-base for Bán's problem. The pattern number is in each case followed by twelve symbols, four for each of the pieces BK, WK and WR. For each piece the first two symbols set the range within which the piece's first co-ordinate (specifying) the east-west bearing) must fall, and the second sets the range for the second co-ordinate. Ranges are ordinarily defined by pairs of integers: UND, MIF and INF denote 'undefined', 'minus infinity' and 'infinite' respectively. Exceptionally the symbol 'P' is used as the first member of a pair to indicate that the second member is a call to a POP-2 truth-function (see text).

Pattern	BK				WK				WR			
BAN0	UND	UND	P	F3	MIF	-3	MIF	-1	MIF	-2	MIF	-1
BAN1	UND	UND	UND	UND	MIF	-3	MIF	-1	MIF	-2	1	1
BAN2	UND	UND	UND	UND	MIF	-3	MIF	2	P	F1	1	1
BAN2A	UND	UND	UND	UND	MIF	-3	MIF	2	2	INF	1	1
BAN3	UND	UND	UND	UND	MIF	-3	2	2	1	1	1	1
BAN4	UND	UND	UND	UND	MIF	-1	1	2	1	1	P	F2
BAN5	UND	UND	UND	UND	MIF	-1	1	2	1	1	2	INF
PW8	1	INF	1	INF	-1	-1	3	3	1	1	3	3
PN4	1	INF	1	INF	0	0	2	2	1	1	2	INF
PN5	1	INF	1	INF	1	1	2	2	2	2	2	INF
PN6	1	INF	1	INF	1	1	2	2	2	2	1	1
PN1	1	INF	1	INF	0	0	2	2	1	1	1	1
PNW2	0	INF	1	INF	1	1	2	2	1	1	1	1
PN3	1	INF	2	INF	1	1	3	3	1	1	2	2
PW3	1	INF	1	INF	2	2	2	2	1	1	2	2
PW4	1	INF	1	INF	2	2	2	2	1	1	1	1
PW5	1	INF	1	INF	2	2	3	3	1	1	2	2
PNC6	1	INF	1	INF	0	0	3	3	1	1	2	INF
PW6	0	0	1	INF	3	3	2	2	2	2	1	1
PNC0	0	0	1	2	2	2	2	2	2	2	1	1
PNC11	0	0	0	INF	1	1	2	2	2	2	1	INF
PWC8	0	0	0	INF	1	1	2	2	2	2	0	0
PNC12	0	0	0	INF	1	1	3	3	2	2	1	INF
PNC10	0	0	1	3	2	2	0	0	1	INF	1	1
PNC1	0	0	1	INF	2	2	2	2	1	1	1	1
PWC4	0	0	0	2	2	2	3	3	1	1	2	2
PWC6	0	0	0	2	2	2	2	2	1	1	2	2
PWC7	0	0	1	INF	2	2	1	1	1	1	1	1

Path-Lengths

It should be noted that m_{WK} is a function of the *sum* of the BK's original co-ordinates. Consequently all BK starting positions which have the same city-block distance from the corner must yield identically the same position at the close of phase 1, and hence the same number of steps in the complete mating path. Thus in the next problem in the sequence illustrated in Figures 6.6a and 6.6b, exactly the same instance of pattern PW8 would be generated (on move 28) whether the BK started at (6,2) or at (5,3). The fact that path-lengths can be expressed as a function of only

TABLE 6.7

Breakdown of the optimal path-lengths of solutions to Bán's problem.
The initial co-ordinates are assumed to be WK : (0, 0); WR : (1, 1); BK : (x, y)
such that $x - y \geqslant 2$; $z = x + y$. $p(z)$ is the 'parity' of z, i.e. 1 if z is odd and 0
if z is even.

		no. of steps in optimal path
part A: outward bound	WR moves:	3
	WK moves:	$z + 3$
	BK moves:	$z + 6$
	total, part A	$2z + 12$
part B: north cycles		$6z - 6$
part C: transition manoeuvre		12
part D: west cycles		$4z + 8 - 4p(z)$
part E: terminal manoeuvre		$7 + 6p(z)$
	total overall	$12z + 33 + 2p(z)$

one parameter, namely the BK's initial co-ordinate-sum z, simplifies the
numerical treatment, which is set out in Table 6.7.

Discussion

The knowledge required to give the optimal solution to Bán's problem on
the infinite one-cornered chessboard was embedded in a data-base comprising only 36 advice triples referencing 28 patterns. Further compression
would be possible, but scarcely profitable without a tighter specification
of what constitutes a pattern. In our case an extra facility – of defining
a co-ordinate class by a POP-2 function in place of an interval bounds-pair
– was introduced so as to be able to break out from the language of
unconditional co-ordinate intervals when required, e.g. for cases where
specification of one co-ordinate requires reference to the other. Motivation
is lacking for a search for principles for forming minimal sets of patterns
in a patched-together description language of this kind. Search for principles with reference to a description language of greater generality is,
on the other hand, a matter of importance to which we intend to return at
a later stage of the investigation.

```
        0  1  2

3     .  .  .

1   BK  .  .

2     .  WR

3     .  WK

WHITE TO PLAY

POSITION MATCHES PNC0

7 MOVES TO MATE.  TARGLIST:(PNC1)
W KING MOVES: NO PLAUSIBLE MOVES.
ROOK MOVES: TRYING WEST
MATCH FOUND WITH TARGET PNC1


        0  1  2

0     .  .  .

1   BK  .  .

2     .  WR  .

3     .  .  WK

BLACK TO PLAY

POSITION MATCHES PNC1

6 MOVES TO MATE.  TARGLIST:(PWC4 PNC2)
B KING MOVES: TRYING NORTH
MATCH FOUND WITH TARGET PWC4


        0  1  2

0   BK  .  .

1     .  .  .

2     .  WR  .

3     .  .  WK

WHITE TO PLAY

POSITION MATCHES PWC4

5 MOVES TO MATE.  TARGLIST:(PWC6)
W KING MOVES: TRYING NORTH
MATCH FOUND WITH TARGET PWC6


        0  1  2

0   BK  .  .

1     .  .  .

2     .  WR  WK

BLACK TO PLAY

POSITION MATCHES PWC6

4 MOVES TO MATE.  TARGLIST:(PWC7)
B KING MOVES: TRYING SOUTH
MATCH FOUND WITH TARGET PWC7
```

```
        0  1  2

3     .  .  .

0     .  .  .

1   BK  .  .

2     .  WR  WK

WHITE TO PLAY

POSITION MATCHES PWC7

3 MOVES TO MATE.  TARGLIST:(PNC10)
W KING MOVES: TRYING NORTH
MATCH FOUND WITH TARGET PNC10


        0  1  2

0     .  .  .

1   BK  .  WK

2     .  WR  .

BLACK TO PLAY

POSITION MATCHES PNC10

2 MOVES TO MATE.  TARGLIST:(PNC11 PWC2)
B KING MOVES: TRYING NORTH
MATCH FOUND WITH TARGET PNC11

        0  1  2

0   BK  .  .

1     .  .  WK

2     .  WR  .

WHITE TO PLAY

POSITION MATCHES PNC11

1 MOVES TO MATE.  TARGLIST:(PWC8)
W KING MOVES: TRYING SW
NO MATCH WITH TARG.
ROOK MOVES: TRYING WEST
MATCH FOUND WITH TARGET PWC8

        0  1  2

3   BK  .  .

1     .  .  WK

2   WR  .  .

BLACK TO PLAY

POSITION MATCHES PWC8

0 MOVES TO MATE.  TARGLIST:(PNC12)
B KING MOVES: TRYING WEST
NO MATCH WITH TARG.

END OF PLAY: TOTAL OF 135 STEPS
```

FIGURE 6.12(a).   Terminal sequences of Figure 6.7(a) run with program parameters set to full output.

```
     0  1  2  3
0    .  BK .  .
1    .  .  . WK
2    .  WR .
```

WHITE TO PLAY

POSITION MATCHES PNC11

7 MOVES TO MATE. TARGLIST:(PWC8)
W KING MOVES: TRYING SW
NO MATCH WITH TARG.
ROOK MOVES: TRYING WEST
MATCH FOUND WITH TARGET PWC8

```
     0  1  2  3
0    .  BK .  .
1    .  .  . WK
2    .  WR .  .
```

BLACK TO PLAY

POSITION MATCHES PWC8

6 MOVES TO MATE. TARGLIST:(PNC12)
B KING MOVES: TRYING WEST
MATCH FOUND WITH TARGET PNC12

```
     0  1  2  3
0    BK .  .  .
1    .  .  . WK
2    .  WR .  .
```

WHITE TO PLAY

POSITION MATCHES PNC12

5 MOVES TO MATE. TARGLIST:(PNC11)
W KING MOVES: TRYING WEST
MATCH FOUND WITH TARGET PNC11

```
     0  1  2
0    BK .  .
1    .  . WK
2    .  WR .
```

BLACK TO PLAY

POSITION MATCHES PNC11

4 MOVES TO MATE. TARGLIST:(PNC10)
B KING MOVES: TRYING SOUTH
MATCH FOUND WITH TARGET PNC10

```
     0  1  2
0    .  .  .
1    BK .  WK
2    .  WR .
```

WHITE TO PLAY

POSITION MATCHES PNC10

3 MOVES TO MATE. TARGLIST:(PNC10)
W KING MOVES: NO PLAUSIBLE MOVES.
ROOK MOVES: TRYING EAST
MATCH FOUND WITH TARGET PNC10

```
     0  1  2
0    .  .  .
1    BK .  WK
2    .  .  WR
```

BLACK TO PLAY

POSITION MATCHES PNC10

2 MOVES TO MATE. TARGLIST:(PNC11 PWC8)
B KING MOVES: TRYING NORTH
MATCH FOUND WITH TARGET PNC11

```
     0  1  2
0    BK .  .
1    .  . WK
2    .  .  WR
```

WHITE TO PLAY

POSITION MATCHES PNC11

1 MOVES TO MATE. TARGLIST:(PWC8)
W KING MOVES: TRYING SW
NO MATCH WITH TARG.
ROOK MOVES: TRYING WEST
MATCH FOUND WITH TARGET PWC8

```
     0  1  2
0    BK .  .
1    .  . WK
2    WR .  .
```

BLACK TO PLAY

POSITION MATCHES PWC8

0 MOVES TO MATE. TARGLIST:(PNC12)
B KING MOVES: TRYING WEST
NO MATCH WITH TARG.

END OF PLAY: TOTAL OF 119 STEPS

FIGURE 6.12(b).   Terminal sequence of Figure 6.6b. Essentially the same sequence as that of Figure 6.6a is entered, but displaced one square to the east. Thus, when the pattern PNC1 is reached (top left), and then PWC8, the BK has flight squares, and six additional steps are required to reach checkmate.

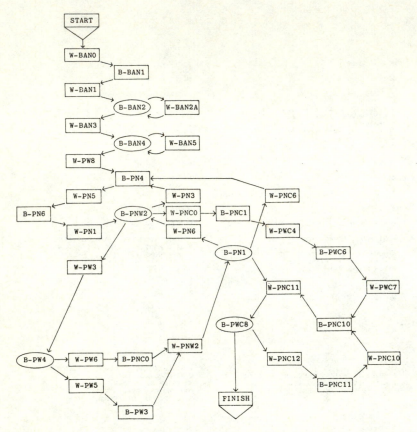

FIGURE 6.13.. Complete strategy for Bán's problem. It divides into parts as follows:

Part A. *Outward bound:* the sequence from W–BAN0 to exit to W–PW8 (goal-patterns 1 and 5 respectively of Table 6.4). The two cycles represent the BK's eastward and southward journeys respectively.

Part B. *North cycle:* this is B–PN4 → W–PN5 → B–PN6 → W–PN1 → B–PNW2 → W–PN3, with B–PNW2 acting as the transfer point for entry to the west cycle.

West cycle: on the first time round this cycle, which starts B–PNW2 → W–PW3 → B–PW4 → W–PW5, Black gains time by playing to W–PNC6 rather than W–PN6. This is referred to in the text as the 'transition manoeuvre'.

Terminal manoeuvre: the end-play follows different routes according to the parity of the starting position. For $z = 5, 7, 9, 11, \ldots$ we have W–PW6 → B–PNC0 → Q–PNW2 → B–PN1 → W–PNC11 → B–PWC8 → W–PNC12 etc. For $z = 6, 8, 10, 12, \ldots$ we have W–PNC0 → B–PNC1 → W–PWC4 → B–PWC6 etc.

Using the crude pattern language of Figures 6.5 and 6.6 it was found easy to make local changes to the strategy, and to detect and mend lapses from optimality during interactive testing of the program. More difficult problems — even KRK on the 8 x 8 board — will need means of automatically generating patterns. The reasons are (a) that constructing patterns from the user's head is burdensome, and in a difficult domain prohibitive. Further, the problem of guarding against human error becomes unmanageable. (b) No proof, only plausible argument, has been given of the assertion that the strategy here exhibited is optimal. For formal proof to be tractable the patterns need to be the product of a fully defined procedure. The guarantee of optimality can then be built into the generating algorithm itself.

Acknowledgements

Part of this work was done during a visit to the Department of Information Sciences, University of California, Santa Cruz, to whom my thanks are due. I also wish to acknowledge programming assistance received during my stay from Mr Michael Lawson.

REFERENCES

Bán, J. (1963). *The Tactics of End-Games* (translated from the Hungarian by J. Bochkor). Corvina Press. Reprinted 1972 by the Budapest Athenaeum Printing House.

Bramer, M.A. (1975). Representation of knowledge for chess endgames. *Technical report*. Milton Keynes, UK: The Open University.

Clarke, M.R.B. (1977). A quantitative study of King and Pawn against King. In *Advances in Computer Chess 1* (ed. M.R.B. Clarke). Edinburgh: Edinburgh University Press.

Fine, R. (1941). *Basic Chess Endings*. New York: David McKay Company, and London: G. Bell and Sons, Ltd.

Huberman, B.J. (1968). A program to play chess end games. *Technical report no. CS 106*. Stanford University: Computer Science Department.

de Latil, P. (1956). *Thinking by Machine* (translated from the French by Y.M. Golla). London: Sidgwick and Jackson.

Nemes, T.N. (1969). *Cybernetic Machines* (translated from the 1962 Hungarian edition by I. Földes). London: Iliffe Books Ltd.

Samuel, A.L. (1959). Some studies in machine learning using the game of checkers. *IBM J. Res. Dev. 3*, 211-29.

Shannon, C. (1950). Programming a computer for playing chess. *Phil. Mag. 41*, 356–75.

Sussman, G.J. (1975). *A computer model of skill acquisition.* Elsevier Computer Science Library, Artificial Intelligence Series No. 1. New York: American Elsevier Publishing Company.

Vigneron, H. (1914). Les automates. *La Natura*, 56–61.

Zermelo, E. (1912). Über eine Anwendung der Mengenlehre auf die Theorie des Schachspiels. *Proc. 5th Int. Cong. Mathematicians*, 501–4. Cambridge. (English translated in *Firbush News 6* (1976). Edinburgh: Machine Intelligence Research Unit.)

Zuidema, C. (1974). Chess, how to program the exceptions? *Afdeling informatica* IW21/74. Amsterdam: Mathematisch Centrum.

APPENDIX

Computer Simulation of Torres y Quevedo's KRK Machine

Specimen play from the Torres strategy (Figure 6.3 of the main text) was obtained by hand-simulation from Table 6.1. As a check, the six decision rules were then re-expressed as an advice-list, shown in Table 6.8, suitable for use with the program. The same sequence of moves was generated by program as had been found by hand. Only White's moves were computed, since Torres' rules only apply to White, and this is reflected in the advice-list. Black's replies were typed into the program using a 'request for suggestions' option, which the program invokes when it cannot find any matching advice for an input position.

Simulating the Torres machine was an interesting supplementary exercise in expressing a strategy as a date file for the advice-taking program. Implementing the strategy itself was quick and easy enough, all the non-Torres-specific work having already been done in writing the main program. On the other hand the limitation of descriptive form to conjunctions of predicates defined on individual co-ordinates was felt to be distinctly cramping. More complex domains than KRK will require less restrictive formats.

TABLE 6.8
Torres' strategy as patterns-and-advice files.

```
TOR0  0 7 0 5 UND UND 2 7 P Q3 UND UND
TOR1  P Q5 0 5 UND UND 2 7 P Q6 UND UND
TOR2  0 7 0 4 UND UND 3 7 UND UND 2 6
TOR3  0 7 0 5 UND UND 3 7 UND UND 1 1
TOR4  0 7 0 5 0 0 2 2 UND UND 1 1
TOR5  0 7 0 5 0 0 2 2 UND UND 0 0
TOR7  0 7 0 5 P Q1 2 7 P Q4 1 1
TOR8  0 7 0 5 P Q2 2 7 UND UND 1 1


Q1:  IS BK FILE DISTANCE ODD?

Q2:  IS BK FILE DISTANCE NON-ZERO AND EVEN?

Q3:  IS PIECE ON SIDE OF THE BOARD?

Q4:  DOES BK FILE DISTANCE EXCEED 1?

Q5  IS PIECE IN ONE OF THE TWO "ZONES"?

Q6:  IS PIECE IN SAME ZONE AS BK?


1 TOR8  0 0 1 0 0 0 0 0 0 0 0 0 0  [TOR7]
1 TOR7  0 0 0 0 0 0 0 0 0 0 1 1  [TOR7]
1 TOR4  0 0 0 0 0 0 0 0 1 0 1 0  [TOR5]
1 TOR3  1 0 0 0 0 0 0 0 0 0 0 0  [TOR3 TOR4 TOR7 TOR8]
1 TOR2  0 0 0 0 0 0 0 0 1 0 0 0  [TOR2 TOR3]
1 TOR1  0 0 0 0 0 0 0 0 0 0 1 1  [TOR0]
```

CHAPTER 7
Chess with Computers

The properties of chess as a domain for problem-solving by machine resemble those of other applications. Much conceptualisation is needed from human practitioners but very little from new computers allowing brute-force searches enormously larger than could be covered by the human brain. Because of its finite and modular structure and the availability of a large corpus of systematised conceptual knowledge bequeathed by the masters, chess constitutes ideal laboratory material for studying the relative effectiveness and computational costs of search-driven *versus* concept-driven approaches.

The history of computer chess begins with Charles Babbage and proceeds through Turing and Shannon to the present day, when tournament programs are beginning to challenge International Masters. In research studies, as opposed to tournament play, the focus has moved to the pattern-based rules mediating Master skill. Programming methods are required which will permit the incorporation of such rules on a large scale.

Understanding the nature, number and means of acquisition of chess patterns is seen as the key. Estimates have been made of the number of chess patterns acquired by a chess-master. Since these are in tens of thousands, accelerated machine uptake by use of inductive generalisation is a necessity. According to initial observations, it is also a practical possibility.

One anticipated outcome of progress along this line is a methodology for using the machine to re-codify, in greatly improved form, human knowledge which at present enjoys only fragmentary, error-infested, inconsistent and unsystematic representations in textbooks. This kind of 'knowledge refining' belongs to the new discipline of Artificial Intelligence.

The Concept of Artificial Intelligence

Artificial Intelligence (AI) workers sometimes complain that they are in a 'no win' situation. 'Would it be intelligent if a machine could read a newspaper and give you a summary of its contents?' enquires the AI scientist. 'Certainly!' concedes his critic. 'My student', replies the AI man, 'has just done that.' 'But how does his program work?' says the critic with an air of suspicion. After a spell with blackboard and terminal he decides that his suspicion was justified. 'So that's all! I don't call that intelligent.'

My AI colleagues find this understandably irksome. Is every implementation of this or that aspect of intelligence to be dismissed the moment the implementation is understood? It is of course agreed that there must be no trickery. A program which prints out just the headlines from a newspaper might achieve a surprisingly respectable summary. But even after tightening up the task definition it might still not require intelligence for a human to perform it. If so, it would be 'Type 0' and declared of no relevance to artificial intelligence research. Simple arithmetic is such a task; searching a dictionary is another.

Relevance begins when the task as defined cannot possibly be done by a human except on the basis of considerable intelligence. AI people have in the past lumped all these tasks together and have said 'If a machine ever does any such thing, then it is intelligent'. It is now becoming apparent that this is wrong, and that the category is compounded of two types. Type 1 demands intelligence from any human solver, but may be machine-soluble by some other route. Type 2 demands intelligence from any solving device whatsoever. In my use of the term 'intelligence' it is intimately related to the memory-saving trick underlying all cognition, namely use of conceptual models.

This is what the critic usually has at the back of his mind: the power of the human brain to conceptualise its task environment. Some tasks he realises, require a high degree of conceptualisation, whether by brains or machines. An example would be watching baseball and preparing a newspaper report of the game. Even annotating a game of chess between strong club players would be regarded as a far more intelligent machine feat than a chess program's exploit in beating the players. The latter exploit would exemplify only Type 1. It certainly demands a high degree of conceptualisation from the human player, as is evident from de Groot's[1] study. But in this era of fast processors and large memories the possibility exists of finding equally good or better solutions by other means. Indeed, chess computer programs already beat strong club players, who rightly regard their vanquishers as unintelligent.

If we redefine the playing task as defeating Fischer or Karpov, then we almost certainly have Type 2. No amount of brute-force computation will then obviate the need for more or less elaborate conceptual representations of chess skill in machine memory. If we further tighten the performance criterion and demand that a correct move be always computed, that is, that the program be error-free in the game-theoretic sense, then we do not know whether the computation is feasible at all on a terrestrial scale, however 'intelligent' the machine. This depends upon the computational complexity of chess, at present unknown.

Consider a domain of great complexity — meteorology. Just as Type 1 tasks can be found, for example error-free play of elementary endings, within the larger Type 2 domain of chess, so Type 1 tasks of weather prediction can be found within the Type 2 domain of meteorology. An example is the preparation of one-week forecasts for the British Isles. It takes years to become a first-class forecaster, and not everybody has the intelligence to become one. But when the European Centre for Medium-range Weather Forecasting gained the benefit last year of an 80-million-instructions-per-second Cray 1 computer, no one doubted the Centre's ability to achieve one-week predictions with little to fear from unaided human competition. Eventually the aspirations of computer-aided forecasting may grow to the point when conceptual and heuristically structured models, of the kind beloved of AI people, will have to be integrated into their present supercalculational representations. By then the machines will have entered Type 2 territory and human forecasters, however intelligent, may be left behind.

Elsewhere[2] I discuss various issues of complexity which bear on what brains and machines can and cannot do. Ideal laboratory material for preparing the way for these topics can be obtained from computer chess, to which the remainder of this paper is devoted.

History of Computer Chess

The first serious proposal to have a machine play chess was made by Charles Babbage, the British pioneer of digital computing in the 19th century. It was never executed. In the early years of this century the Spanish engineer Torres y Quevedo demonstrated an electro-mechanical device for mating with king and rook against king. The first experiments with the complete game were conducted in the late 1940's by the British logician A.M. Turing.

According to the testimony of I.J. Good,[3] interest in the idea of mechanising the play of intellectually challenging games had taken root in Turing's mind as early as 1941. After the war he collaborated with D.G. Champernowne and others to construct various paper machines embodying mechanized strategies for chess. Play was poor.

Of Turing's own chess prowess, International Master Harry Golombek[4] wrote:

. . . His marvellously inventive mind constantly delighted me but, curiously enough, though passionately keen on chess, he was a weak player; so weak that, upon his resigning to me, I was able to turn the board round and win against him with the colour with which he had resigned.

Turing[5] refers to experimental tests. These were under way already in 1947, when he and Champernowne engaged with S. Wylie and myself[6] in an uncompleted paper machine experiment. He made subsequent attempts to complete the experiment by writing a program for the Ferranti Mark 1 machine at Manchester, and at the time of his death in 1954 he had active plans to continue the work.

In 1950 Claude Shannon published a paper[7] which went beyond Turing's proposals in important respects, though few of these extensions have yet been implemented. The Turing–Shannon 'lookahead–evaluate–minimax' paradigm still forms the basis of today's tournament program.

During the same period de Groot's[1] study of the thought-processes of chess-masters revealed that their special ability does not reside in computer-like qualities of memory or of accurate, fast and sustained calculation, but from powers of conceptualisation. This result had been foreshadowed by Alfred Binet's[8] investigation in 1894 of the ability of chess-masters to play many games of blindfold chess simultaneously. Binet concluded that in addition to *la mémoire* this accomplishment rested on *l'érudition*, the use of accumulated chess knowledge to form meaningful descriptions of board positions, and *l'imagination*, the ability to reconstruct mentally a position from a description. By *la mémoire* Binet did not intend the kind of photographic representation postulated in so-called 'eidetic' imagery. The point receives emphasis from Reuben Fine's[9] introspective memoir on blindfold play: 'It seems to be impossible to think of the board without relationship to the pieces, and the pieces without relationship to the board' and '. . . this ability is related to the more general process of concept formation . . .'. It seems that the chief significance of *la mémoire* lies in its use to give effect to the other two capabilities. Although machine play is now impressive, little progress has been made in mechanising *l'érudition* and *l'imagination*.

The earliest chess programs were developed in the 1950s and have been reviewed by Samuel.[10] The modern era dates from the 1967 entry into human tournaments of the Greenblatt–Eastlake–Crocker[11] program. Play was at the level of a weak-to-middling club player, Class 'C', US Chess Federation (USCF) rating about 1500. Expert level begins at about USCF 2000, National Master at 2200, International Master at 2400, Grandmaster at 2500.

The first World Computer Chess Championship,[12] held in 1974 was won by the program KAISSA developed by V.L. Arlazarov, G.M. Adelson-Velskiy, A.R. Bitman and M.V. Donskoy. Standards of play corresponded approximately to USCF 1600. In 1977 the second World Championship was won by the program CHESS 4.6 of L. Atkin and D. Slate, with KAISSA second. Play rated about USCF 2000. Program improvements contributed to progress, but the major factor was order-of-magnitude increases in speeds of computing hardware to 15 million per second and 3 million instructions respectively.

Further large speed-ups are now available through the development of special-purpose chess circuitry, as in the CHEOPS device completed by Greenblatt and Missouris[13] at MIT, and the appearance of new main frames such as the Cray computer mentioned earlier. Tenders placed by American institutions for machines of highly parallel architecture with over 10^9 instructions per second capacity (NASA Ames Center), and on-line memories of 10^{16} bits (Lawrence Radiation Laboratory), add further indication that the next few years' hardware development may alone suffice to force the door to Master-level play. However, for scientific study of cognitive science and cognitive engineering, more ambitious goals of performance, or of computational economy, must be set.

Scientific Aspects

Computer chess has been described as the *Drosophila melanogaster* of machine intelligence. Just as Thomas Hunt Morgan and his colleagues were able to exploit the special limitations and conveniences of the *Drosophila* fruit-fly to develop a methodology of genetic mapping, so the game of chess holds special interest for studying the representation and measurement of knowledge in machines. Its chief advantages are: chess constitutes a well-defined and formalised domain; it challenges the highest levels of intellectual capacity over a wide range of cognitive functions — logical calculation, rote-learning, concept-formation, analogical thinking, deductive and inductive reasoning, and so forth; a detailed corpus of chess knowledge has accumulated over centuries in chess instructional works and commentaries; a generally accepted numerical scale for performance is available in the USCF rating system; and finally, the game can readily be decomposed into sub-games which can be subjected to intensive separate analysis.

To clear the ground for further discussion of the knowledge approach, we must first review the attainment level and mechanisms of state-of-the-art chess programs and the reader should learn about a recent performance of CHESS 4.6 (see Appendix).

State-of-the-Art Mechanisms

The fundamental Turing–Shannon machine procedure involves looking ahead from the current position along a branching tree of possibilities. Human players also look ahead, through concrete analysis, but on a severely restricted scale. According to de Groot, 30 positions mark the limit to what a Master normally holds in his lookahead memory, and there is general agreement that the upper bound may safely be placed at about one hundred positions. By contrast, chess programs commonly grow lookahead trees in computer memory comprising hundreds of thousands of positions.

If no rules for pruning unpromising branches were applied, branching of the lookahead tree would be around 30 branches per position in the mid-game. Without such pruning the search would accordingly grow almost by powers of 30 (the 'almost' is for the occurrence of confluences in the lookahead tree which could be detected and eliminated at some computational cost), and would soon be defeated by what has been termed the 'combinatorial explosion'. Pruning rules are of two types.

The first type, to which the term 'alpha-beta pruning' has been applied, preserves completeness. That is, it is guaranteed not to affect the move-choice finally extracted from the search. In favourable cases such pruning doubles the number of ply over which search can be made for a given number of positions examined in lookahead. (A ply is a half-move. A move usually denotes a complete 'White moves, Black moves' cycle.) A simple illustration of the rule's operation is shown in Figure 7.1. The animating principle is that once evidence is to hand that a variation is not the best, no further computational work need be expended on determining by just how far it falls short.

The second type of pruning, sometimes called 'forward' or 'plausibility' pruning, suppresses branches when features detected in the position from which branches would be grown indicate that they are likely to lead to bad continuations — for example, involving uncompensated loss of material. Forward pruning heuristics are many and various, and carry the risk of eliminating a line which continued search would have picked as the best. The strongest contemporary program, CHESS 4.6, makes no use of forward pruning.

All programs apply termination rules to halt growth of the tree beyond certain limits. The main factor in termination is the occurrence of quiescent positions, Turing's 'dead' positions, in which no captures or forced moves are in immediate prospect.

In the Turing–Shannon scheme the program applies an evaluation function to the terminal positions of the lookahead tree, labelling them with computed estimates of their strategic strength or weakness. The leading term of the function is derived from a count of own and enemy pieces assigning weights such as the conventional valuations: 9 for queen, 5 for rook, 3 for bishop, 3 for knight, 1 for pawn. Bonuses may be added for such qualifying factors as the two-bishop advantage, or the advancement of pawns. Mobility normally comes next. Its importance was established empirically by the British clinical psychologist Eliot Slater[14] in 1950. His results are shown in Table 7.1. Other features contributing terms to the evaluation function are king safety, control of central squares or squares in the neighbourhood of the enemy king, pawn structure, rook control of files, and so forth. A tournament program's evaluation function

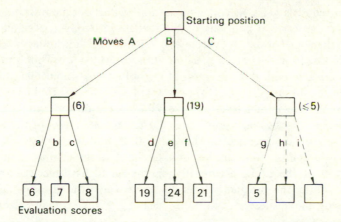

FIGURE 7.1. 'Toy-sized' search tree for illustration of the basic idea of alpha-beta pruning. The search here goes to level 2 and is conducted 'depth first'. Scores which have been backed up by the minimax rule are shown in parentheses and broken lines denote a part of the tree which the alpha-beta rule eliminates as unnecessary to explore. Such exploration could only alter downwards the score provisionally associated with the right-hand level-1 position, so that move B would be selected in any case. Move g is called a *refutation move* to move C.

Because the tree drawn here is only two-ply deep and the first refutation move was encountered late in the search, the economy appears small in this diagram. In the general case, when the rule is applied recursively level by level, savings are very significant. In the ideal case the number of positions requiring to be examined is reduced to the square root of the full number. Figure 7.2 shows a slightly larger case, illustrating the minimax procedure on which alpha-beta operates.

TABLE 7.1

Results of an examination by E.T.O. Slater of 78 arbitrarily selected master games which ended with a decisive result on or before the 40th move.

After move	Winner's mobility (average)	Loser's mobility (average)	Difference
0	20.0	20.0	0
5	34.2	33.9	0.3
10	37.5	36.0	1.5
15	39.7	35.2	4.5
20	38.9	36.4	2.5
25	39.6	31.9	7.7
30	35.6	27.7	7.9
35	31.7	23.2	8.5

typically consists of a weighted combination of thirty or more such terms.

After the terminal positions of the lookahead tree have been allotted values, these are backed up the tree by a minimax rule illustrated in Figure 7.2. Starting from the bottom, each position for which it is White's turn to play is credited with the *maximum* of the values of its immediate successors, on the presumption that White wants to get the best for himself; the convention here is that large positive values are good for White, bad for Black. Each Black-to-play position is allotted the *minimum* of the values of its immediate successors. The wave thus spreads through the tree from the terminal positions towards the root, until all the immediate successors of the position currently on the board have received values. Playing White, the program selects the move which will give the highest-valued of these successor positions. Playing Black, it will select the lowest.

Various devices for eliminating redundant calculation and redundant storage can be used to improve this basic mechanism. The most far-reaching is KAISSA's use of a theory of influence to avoid repeated computation of sub-trees identical in all but inessentials. Allowing for overheads incurred in applying the method, the saving is approximately twofold.

The depth to which modern computer chess programs look ahead ranges over 6–15 ply, somewhat in excess of the typical depth of analysis by Masters and Grandmasters as measured by de Groot. Correspondingly, in sharp tactical situations machine play is now comparable with Grand-master play, while meriting no more than 'C' rating for positional play. Machine tactical analysis is now so strong that decisions may sometimes be obscure to human onlookers even though soundly based. A case in point arose in the 1977 World Championship in KAISSA's losing game against Duke University's program DUCHESS, as shown in Figure 7.3.

Limitations of the Turing–Shannon Paradigm

The standard of overall play has yet to attain solid Master level. Some of the reasons are as follows.

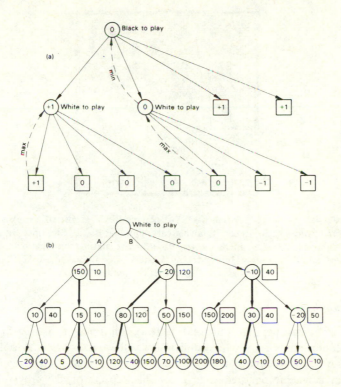

FIGURE 7.2(a).   The root of this two-level lookahead tree acquires a value by alternate application of the 'max' and 'min' functions. If alternation is extended backwards from all terminal positions of the game tree, the initial position of the entire game will ultimately be assigned a value. Terminal positions are shown as boxes.

(b) Lookahead tree in which the positions are marked with 'face values' assigned by an evaluation function (bars over negative values). Boxed values are values backed up by minimax from the lookahead horizon. Arrows with broken lines correspond to moves which would not need to be explored under the alpha-beta pruning regime: the move finally selected by White is B, as having the largest of the three backed-up values. Why not A, as having the largest of the three face values? Curiously enough, although supported both by intuition and by experiment, the rationale for the back-up basis of move selection remains to be formally explicated.

FIGURE 7.3.   In this position KAISSA, playing Black, sacrificed a rook with 34.
. . . Re7–e8. Although this clever move greatly prolongs Black's life, since the alternative of moving the king opens the way to a 5-move mating combination by White,
500 spectators, including International Masters and a former World Champion,
assumed that it was a blunder.

Horizon Effect

Berliner[15] has pointed out that reliance on unaided Turing–Shannon
procedures renders a program oblivious to all events beyond its lookahead
horizon. Even though a trans-horizon loss, or a trans-horizon gain, may
appear inevitable or obvious to the human spectator, the program plans
from hand to mouth, short-sightedly sacrificing material to delay a loss
which cannot indefinitely be averted. Alternatively it may forfeit an
eventual large expectation by grabbing at an immediate small gain.

Here is a case as reported in *Computer Weekly* of 20 April 1978 from
play against CHESS 4.6 at the Aaronson European Chess Congress, 1977.
Mr L. Perry had the White pieces and play went as follows (algebraic
notation):

English opening, transposing to Queen's gambit accepted.

L. Perry	CHESS 4.6
White	Black
1. c4	Nf6
2. Nc3	d5
3. d4	d x c4

Black: Chess 4.6

White: L. Perry
Position after 13. b × c3

FIGURE 7.4.

4.	Nf3	a6
5.	e4	b5
6.	a4	b4
7.	Na2	N x e4
8.	N x b4	e5
9.	Nc2	e x d4
10.	Q x d4	Q x d4
11.	Nf x d4	c3
12.	Bc4	Nd7
13.	b x c3	

At this point CHESS 4.6's seven-ply lookahead found a seven-ply combination forcing the gain of a pawn. But the program could not detect within this horizon that the spot in which Black's knight ends up is a death-trap. The next few moves went:

13.	...	Ne5
14.	Bd5	Nd3 +

15.	Kf1	Ne x f2
16.	B x a8	N x h1,

stranding the knight.

It might be asked whether the fateful Ne5 was necessarily based on lookahead of just seven-ply. Does not CHESS 4.6 sometimes analyse more deeply? Yes, but only along forcing sequences of captures and checks. The horizon position created by 16 . . . N x h1 is quiet. Another 15 moves elapsed in the game before the knight was finally taken. But its role had been reduced to that of a passive onlooker. Black resigned on the 33rd move, almost immediately following its capture.

The human player sees that the knight will be trapped. He does this not by lengthy and detailed lookahead analysis but by perceiving familiar patterns remembered from past experience. Pattern-knowledge of this kind is fundamental to chess, as indeed it is to cognition generally.

It is not to be supposed that the human's more far-sighted awareness is based on a longer lookahead in the ordinary sense of detailed analysis. We have already seen that the balance of tactical analysing power lies, these days, in the other direction. As is evident from a careful scrutiny of the examples in Figure 7.5, the longer view is derived from processes of a different kind, involving reasoning about the broad properties of the position concerned. This type of 'conceptual lookahead' discovers, for example, that the bishop in Figure 7.5(a) is lost without need to follow all the forward variations. Powers of abstract and causal reasoning have been implemented by Berliner in experimental laboratory studies, and a number of workers are studying the incorporation of chess pattern-knowledge into computer programs. Tournament programs, however, so far remain devoid of these capabilities.

Lack of Long-range Strategic Ideas

A Master plans at the conceptual level, linking the main milestones with detailed steps in a separate operation. Contemporary programs have no corresponding capability. In the end-game in particular, where long-range reasoning is at a premium, computer programs can flounder aimlessly promoting small disconnected goals with no unifying strategic thread.

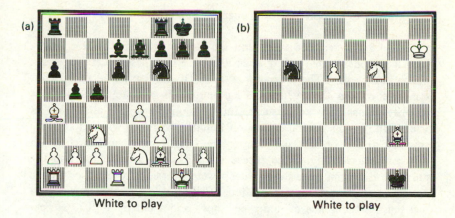

FIGURE 7.5.   (a) With a five-ply lookahead White appears to escape loss of the bishop by pushing the pawn. By the time Black has replied P x P in this branch of the lookahead the bishop loss lies on the far side of the horizon. (b) Greed in going immediately for pawn promotion is rewarded by the tedium of playing out a bishop and knight mating sequence. Since the enemy knight is tied down, the bishop can move round to capture him at leisure.

Lack of Locally Applicable Heuristics

By themselves advances in computing power are not expected radically to improve the situation described above. Remedies must take their departure from an appreciation of the chess master's ability to utilise very large bodies of conceptualised and indexed knowledge. In programs of the Turing–Shannon type, the sole significant repository of chess knowledge is the evaluation function, which embodies only those heuristic principles which have uniform application over the myriad sub-games which constitute the complete game of chess. Principles which apply in some kinds of situation, but not in others, cannot easily or compactly be accompanied by such a structure. The point can be illustrated by the results of submitting to CHESS 4.5, an almost identical earlier version of CHESS 4.6, three test positions from the ending king and knight against king and rook (KNKR).[16]

Performance of CHESS 4.5 Tournament Program with KNKR

CHESS 4.5 was required to defend the weaker side of KNKR against a human opponent rated just over 2000 on the US Chess Federation scale, that is, an 'expert'. The program ran on a CDC 6400 machine, on which it was able to win the 1976 ACM Computer Chess Championship; more recently it has had highly successful trials on the much faster CYBER 176. CHESS 4.5's general evaluation function was used without allowing any adjustment or special tuning to the KNKR problem. Search depth was set to seven ply. Since forced variations are searched beyond this preset horizon, moves eight ply deep were occasionally searched in the present case. Under these conditions, CHESS 4.5 typically looked at a few tens of thousands of nodes per move and spent up to 120 sec per move, typically between 30 and 60 sec. The three trials are described below.

1) A classical difficult defence described by Fine[17] and Keres[18] with the weaker side's king in the corner. CHESS 4.5 found correctly the move considered most difficult in the books, but then stumbled on the fourth move of the main 'book' variation, obtaining a lost position. Starting from the position of Figure 7.6(a) with CHESS 4.5 as Black, the play went:

1.	Rb2+	Ka1
2.	RB8	Ne2! (The move considered to be most difficult)
3.	Kb3	Kb1
4.	Re8	Nd4 + ? (Keres gives 4 . . . Nc1 +)
5.	Kc3	Nb5 + (If 5. . . . Nf3, 6. Re3 separating K and N definitely)

Play actually stopped here, but Black is lost as the further analysis shows: 6. Kb4 Nd4, 7. Re4 Nc2 +, 8. Kc3 Na3 (If 8 . . . Kc1, 9. Re1 + Ka2, 10. Re2 + Kb1, 11. Kb3, and White wins).

2) Figure 7.6(b) shows another difficult position, with the weaker side's king on the edge. CHESS 4.5 playing Black found the only correct defence against the main line given by Keres (exclamation marks by Keres):

1.	Rb7	Nh6
2.	Rh7	Ng8!

In a game Steinitz–Neumann (1890, with colours reversed), Black wrongly continued 2. . . . Ng4 and lost after 3. Rh4 Ne3, 4. Re4! Nd1, 5. RF4 + Kg7, 6. RF3 Kg6, 7. Ke5 Kg5, 8. Kd4 Kg4, 9. Rf1 Nb2, 10. Rb1 Na4, 11. Rb4 winning the knight.

3.	Rf7 +	Ke8
4.	Rg7	Kf8!
5.	Rh7	Ke8
6.	Rf7	Nh6
7.	Rf1	Ng8

This preserved the draw. Or, as also tried, instead of 7. Rf1:

7.	Rg7	Kf8

3) A further position (Figure 7.6(c)) has the weaker side's king in the centre, the easiest defence. CHESS 4.5 playing Black allowed its king to be driven to the edge resulting in a harder defence. This enabled the opponent to create mating threats, and after additional weaker moves by the program the king and knight became separated, leading to a lost position. In the game-record below, numerals in parentheses give the number of moves to knight-capture given best play. '?' denotes a mistake, that is, a White move which turns a win into a draw or a Black move which turns a draw into a lost position. '?' in parentheses denotes an inaccuracy by White, that is, a move which turns a position winnable in a few moves into a position requiring many moves to win.

1.	. . .	Kf5	
2.	Kc4	Nh4	
3.	Kd4	Kf4	
4.	Rf2 +	Nf3 +	
5.	Kd3	Kg3	
6.	Ke3	Ne5	
7.	Rf8	Kg4	
8.	Ke4	Nc4?	(After this Black is already lost; better is 8. . . . Ng/6)

9.	Rg8 + (10)	Kh5	(Now king and knight are separated)
10.	Kd4?		(10. Rd8! wins)
		Nd6	(Drawn)
11.	Ke5	Nc4 + ?	(Does not care about joining king and knight; Nf7 + draws)
12.	Kf4 (10)	Nb2 (8)	
13.	Rg5 + (8)	Kh6 (7)	
14.	Kf5 (11) (?)		(RB5! wins easily)
		Nc4 (11)	
15.	Kf6 (22) (?)	Ne3 (22)	
16.	Ra5?	Ng4 +	(Drawn)
17.	Kf7	Nh2	
18.	Ra1	Ng4	

The position is drawn again and play was terminated. Although starting from an easy drawn position, the program, lacking fundamental knowledge about KNKR endings, was wandering from one lost position to another. This last example shows that a sneaky player can probably, starting from any KNKR position, always induce the program to disjoin king and knight and get into a lost position. The program's 'global' evaluation function sees no reason why king and knight should not be separated!

It is interesting to observe that the program's opponent, although an expert, after achieving theoretically won positions never grasped the opportunity actually to defeat the program. This has a bearing on the level of difficulty of this subdomain.

Chess Patterns and their Acquisition

The underlying cause of CHESS 4.5's failure on the above-described test is the unsuitability of a unitary 'global' evaluation function to represent the fine structure of chess knowledge as we find it in standard expository works such as Reuben Fine's *Basic Chess Endings*. Tan[19] has remarked that chess 'may be considered as a mosaic of thousands of sub-problems, each requiring a different way of handling'. Contemporary research aims beyond the Turing–Shannon paradigm at the means for constructing specialised knowledge-sources for each of the thousands of sub-problems of the mosaic.

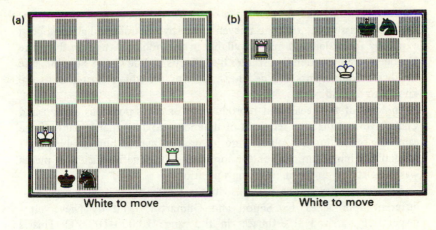

FIGURE 7.6.   Three starting positions used for trials of the CHESS 4.6 computer program, as described in the text.

Illustrative of this style is Tan's own formalisation of concepts for characterising pawn structures. His program describes pawns-only positions in terms of 'islands' and 'fronts', forming pawn-relation graphs from which 'attack–defence diagrams' may be constructed. The aim is to generate an automatic summary of the position's dynamic potentialities.

Tan obtained his primitive relations from established formulations of chess theorists — hostile relations such as counterpawn and sentry, and friendly relations like duo and protector. In more complex subdomains there is no assurance that the relevant primitive notions will all have been explicitly identified and made available in the books. Attention therefore has been given to extending studies like Binet's and de Groot's to elucidate the basic components of the chess patterns on which Master skills are known to depend.

In tests of the recall and reproduction of chess positions. Simon and Chase[20] found that the relations of defence, proximity, and being of the same denomination or colour were all used as mental building-blocks. A particularly important relation for composing meaningful clusters of pieces was joint attack on the enemy king's position.

How many descriptive patterns does a Master store in memory? Two different approaches, by Simon and Gilmartin[21] and by Nievergelt[22] respectively, arrived at estimates in the range 10,000–100,000. This is comparable with human pattern-capabilities in some other contexts. There are, for example, approximately 50,000 Chinese ideograms.

The labour of introducing so large a number of chess patterns into computer memory would clearly be prohibitive if each pattern had to be specified in detail, with all its quirks and exceptions, by conventional programming. The labour could be eased if the machine system were capable of inductively acquiring patterns from playing experience or from user-supplied tutorial examples. This has not yet been achieved for any program, and chess systems capable of handling patterns and associated heuristic rules acquired from a human tutor are in their infancy, as Bratko, Kopec and Michie[23] reported. Inductive learning of patterns has been shown feasible for the limited case of the king and pawn *versus* king ending by Michalski and Negri.[24,25] It is not yet clear to what extent the methods used are capable of generalisation to larger and more complex subdomains.

A Pattern-based Program

A computer program into which a chess-master's pattern-knowledge can be incorporated manifests a new phenomenon. The machine is able to give explanations of its decisions expressed in concepts familiar to human players. The records reproduced earlier from trials of CHESS 4.6 with the king-knight-king-rook end-game represented everything humanly digestible which the machine was capable of reporting. It would have been possible to arrange additionally for output of the internal analysis tree underlying each move-choice; but since such trees typically consist of many tens of thousands of nodes, such an 'introspective record' would be of little value.

Knowledge-based programs studied in work on machine intelligence differ conspicuously in this respect. To develop such programs in general requires special computer languages, tuned to the expression of descriptive concepts rather than imperative commands. Using the AL1 language developed at Edinburgh, Ivan Bratko embodied the master knowledge required for the defence of king and knight against king and rook in an Advice Table of the language. The following record was obtained when the system was given a position, theoretically lost in 27 moves, and required to defend it against a chess-master of USCF rating exceeding 2300. As might be expected of a machine representation specialised in this one particular end-game, it conducted the defence a great deal more skilfully than did CHESS 4.6. More significant was its ability to give an account of its thought-processes in terms comprehensible to others. Herein lay the crucial contrast between the new knowledge-intensive programs and the compact, knowledge-poor programs of the computer science mainstream tradition.

Two informal trials were made of the master's ability to win against the program, starting from a theoretically won, but difficult, position. For each trial the chess-master was given the following starting position. It is the start of one of the two longest wins in the KRKN domain discovered by Thomas Strohlein[26] from an exhaustive computation of the complete problem space of approximately 3 million positions.

Game 1

1.	Rh4!		(Optimal move in the minimax sense)
1.	...	Nf6?!	(Suboptimal defence: the minimax path length is now 17 moves; optimal was 1. ... Ne5)
2.	Kd2	Kb2	(Both moves optimal)
3.	Rd4?	Kb3	(After this the position is drawn; correct was 3. Kd3)
4.	Kd3	Ne8	

White to move

White can mate or win the knight
in 27 moves against best defence

5.	Rd7	Kb4	
6.	Kd4	Kb5	
7.	Re7	Nd6	
8.	Kd5	Kc?	= Draw agreed.

After a short post-mortem and analysis, the master tried again:

Game 2 (starting from same position as in Game 1)

1.	Rh4!	Nf6?!	
2.	Kd2	Kb2	
3.	Rf4?		(This mistake brings the play from a 16-move win into a draw)
3.	...	Nd7?	(Correct was Nd5. Position is now lost in 20 moves)
4.	Kd3?		(Now position is drawn again. Winning line is 4. Rb4+ Ka3, 5. Kc3 with a win in 19 moves)
4.	...	Kb3	
5.	Kd4	Kb4	= Draw agreed

The comments made above are those of the Advice Table's author, Mr I. Bratko, a master-strength player. Below are the machine's own comments in the form of the 'motivations' of its moves.

Play was guided according to principles described elsewhere[16] by four self-administered 'pieces-of-advice' selected by the program from an 'advice-library' to meet each board situation as it arose. They were as follows:

Source of guidance for constructing strategy-tree: Advice 'HOLD 1'.
Phase of play: Game 1, move 8.
Content of advice: Try to achieve a depth-4 position against all possible opponent's replies in which the opponent cannot win our knight on his next move. Our king is not to be left closer to an edge or corner than before. Our king is to be kept throughout at least as close to their king as our knight is, while avoiding a knight's-move relation between our king and our knight.

Source of guidance: Advice 'APPR-0'.
Phase of play: Game 1, moves 1, 2, 5, 6, 7; game 2, moves 1, 2, 4, 5.
Content of advice: Try out successor-positions in forward search according to the following criteria: prefer knight-moves towards our king and away from the enemy king, with a bias towards centralising the knight. Lines may be investigated to a depth of 6, but at depth 4 consider only checking and attacking moves, and at depth 2 investigate only threats to our king or knight and moves that constrain the mobility of our knight. Find a strategy which either captures the enemy rook or, failing that, will avoid being mated, or losing or endangering or trapping our knight, while avoiding putting our knight into the corner or into the special pattern 'edge-loss', and not allowing it to drift away from our king or worsening in any other way the preference criteria given earlier.

Source of guidance: Advice 'APPR-1'.
Phase of play: Game 1, move 3.
Contact of advice: Same as 'APPR-0' but dropping the embargo on allowing our knight to drift away from our king.

Source of guidance: Advice 'APPR-2'.
Phase of play: Game 1, move 4; game 2, move 3.

Content of advice: Same as 'APPR-1', but allowing investigation of lines which temporarily worsen the preference criteria, provided eventual restoration can be seen.

Concluding Remarks

Current endeavours centre round four foci:

1) design of data-structures in forms not only suitable for representing conceptualised knowledge — descriptions, patterns, theories — but also convenient for the user to modify and increment interactively;

2) incorporation into the program of powers of inductive inference, so as to be able to acquire new knowledge from tutorial examples, and also from internal computations;

3) development of conceptual interfaces between program and human expert, facilitating both 'teaching' the computer program and testing what is taught;

4) reversal of the direction of teaching and testing. The use of chess in computer-aided instruction experiments deserves study.

Relevant in this last connection is a finding reported by Bratko, Kopec and Michie.[23] After construction and validation of a suitable knowledge-representation for play of a given ending, this representation was translated back into English. The resulting instructional text was found to be more concise, more accurate, and easier to memorise and use than the master texts from which the knowledge had originally been obtained.

REFERENCES

1. de Groot, A. (1965). *Thought and Choice in Chess*, (ed. G.W. Baylor). The Hague and Paris: Mouton (translation, with additions, of Dutch version of 1946).
2. Michie, D. (1977). Practical limits to computation. *Research Memorandum MIP-R-116*. Edinburgh: Machine Intelligence Research Unit.
3. Cited in B. Randell's 'Colossus', *IEE Trans.*, forthcoming.
4. Golombek, H. 7 November 1976. *Observer Magazine*.
5. Turing, A.M. (1953). Digital computers applied to games. In *Faster than Thought*, (ed. B.V. Bowden), pp. 286–310. London: Pitman.

6. Maynard Smith, J. and Michie, D. (1961). Machines that play games. *New Scientist*, **12**, 367–369.

7. Shannon, C.E. (1950). Programming a computer for playing chess. *Philos. Mag.* 7th Ser., **41**, 256–275.

8. Binet, A. (1894). *Psychologie des grands calculateurs et des jouerus d'échecs.* Paris: Hatchette.

9. Fine, R. (1965). The psychology of blindfold chess: an introspective account. *Acta Psychol.* **24**, 352–370.

10. Samuel, A.L. (1960). Programming computers to play games. In *Advances in Computers*, Vol. 1, pp. 165–192. London: Academic Press.

11. Greenblatt, R.D., Eastlake, D.E. and Crocker, S.D. (1967). The Greenblatt chess program. *Proc. FJCC*, pp. 801–810.

12. Hayes, J.E. and Levy, D.N.L. (1976). *The World Computer Chess Championship.* Edinburgh: Edinburgh University Press.

13. Moussouris, J., Holloway, J. and Geenblatt, R.D. (1979). CHEOPS: a chess-oriented processing system. In *Machine Intelligence 9*, ed. J.E. Hayes, D. Michie and L.I. Mikulich, in press. Chichester, Ellis Horwood; New York: John Wiley.

14. Slater, E.T.O. (1950). Statistics for the chess computer and the factor of mobility. In *Proceedings of the Symposium on Information Theory*, pp. 150–152. London: Ministry of Supply.

15. Berliner, H.J. (1974). Chess as problem solving: the development of a tactics analyser. *Ph.D. Dissertation*, Pittsburgh: Carnegie–Mellon University.

16. Michie, D. and Bratko, I. (1978). Advice Table representations of Chess end-game knowledge. In *Proceedings of IIIrd AISB GI Conference on Artificial Intelligence*, (ed. D. Sleeman) pp. 194–200.

17. Fine, R. (1941). *Basic Chess Endings.* New York: David McKay Co.

18. Keres, R. (1974). *Practical Chess Endings.* London: Batsford.

19. Tan, S.T. (1977). Describing pawn structures. In *Advances in Computer Chess 1*, (ed. M.R.B. Clarke) pp. 74–88. Edinburgh: Edinburgh University Press.

20. Simon, H.A. and Chase, W.G. (1973). Perception in chess. *Cogn. Psychol.* **4**, pp. 55–81.

21. Simon, H.A. and Gilmartin, K. (1973). A simulation of memory for chess positions. *Cogn. Psychol.* **5**, 29–46.

22. Nievergelt, J. (1977). Information content of chess positions: implications for chess-specific knowledge of chess players. *SIGART Newsl*, **62**, pp. 13–15.

23. Bratko, I., Kopec, D. and Michie, D. (1978). Pattern-based representation of chess end-game knowledge. *Computer Journal* **21**, pp. 149–153.

24. Michalski, R.S. and Negri, P.G. (1977). An experiment on inductive learning in chess end games. In *Machine Intelligence 8*, (eds. E.W. Elcock and D. Michie) pp. 175–192. Chichester: Ellis Horwood. New York: John Wiley.

25. Negri, P.C. (1977). Inductive learning in a hierarchical model for representing knowledge in chess end games. In *Machine Intelligence 8*, (ed. E.W. Elcock and D. Michie) pp. 193–206. Chichester: Ellis Horwood. New York: John Wiley.

26. Strohlein, T. (1970). Untersuchungen über kombinatorische Spiele. Dissertation for Dr. Rernat. Munich: Technische Hochschule.

APPENDIX

Performance by CHESS 4.6

The winner of the 1977 World Computer Chess Championship, CHESS 4.6, was given a board against the US Champion Grandmaster Walter Browne in a 44-board simultaneous display at Minneapolis the following year. Such a display imposes a severe handicap, so that Browne's unexpected loss to the machine occasioned surprise but not incredulity, in the chess world. The machine's play showed interesting features which are discussed below by US National Master Danny Kopec in a commentary on the full game record.

The algebraic notation is used for describing the game, as elsewhere in this paper. For this, the ranks (horizontal rows) of the chess-board are numbered from 1 to 8 and the files (vertical columns) are labelled a, b, c, . . . , h, as has been done on the diagram of Figure 8.3. The move denoted 'Re7–e8' in the figure's legend means the 'rook on square e7 moves to the square e8'. If the target square were already occupied by an enemy piece then the symbol 'x' would be used in place of '–', thus indicating that the move involved a capture. For pawn moves the piece symbol is omitted, so that 'e2–e4' means 'the pawn on e2 moves to e4'. The symbol '+' denotes a checking move, that is, a move which threatens immediate capture of the enemy king. The symbols 'O–O' and 'O–O–O' mean 'castles on the king's side' and 'castles on the queen's side' respectively. In the common form of algebraic notation, designation of the source square is omitted, and also the '–' symbol. The three moves hypothesised above then become shortened to 'Re8', R x e8' and 'e4'. 'Rad8' means 'the rook on file a moves to d8'.

Mr Kopec's Commentary

In Spring 1978, three times consecutively US Champion, Grandmaster Walter Browne, gave his second nationwise exhibition tour. Before stopping at The Twin Cities, Minnesota, Browne had accumulated an amazing record of only two losses and six draws in seventeen exhibitions. Due to its 5–0 sweep of The Twin Cities Open, the World Computer Chess Champion, Northwestern's CHESS 4.6 (written by Larry Atkin and David Slate), running on a Control Data CYBER 176, was allowed to be one of Browne's 44 opponents.

English Opening (by transposition)
White: Walter Browne (USCF rating 2560)
Black: CHESS 4.6/CYBER 176 (USCF rating 2070)

1.	d4	Nf6
2.	c4	c5
3.	Nf3	cd
4.	Nxd4	e5

For this event CHESS 4.6 had been modified to 'think' during all the time it took Browne to get around to all 44 boards. This time, though certainly quite variable throughout the entire exhibition, was only a few minutes for the first six times around.

Programs are now strong enough to be able to make real contributions to chess opening theory, and not just dubious, though interesting, moves. The move played, though not new, indicates a tendency for sharp play.

5.	Nb5	Bc5	The point of Black's previous move and it means that CHESS 4.6 would have met 6. Nd6+ with . . . Ke7! and not 6. . . . Bxd6 because the same position a tempo up after 5. . . . Nc6 could have been related. After 6. Nd6+, Ke7, 7. Nf5+, Kf8 Black would stand well thanks to his fine development and the loose White knight on f5 which has already moved 5 times! In this opening, White can never play Bg5 because of . . . Bxf2+ and . . . Ne4+.
6.	N1 c3		This move, played with hardly any hesitation, takes CHESS 4.6 out of its 'book'. It takes 2 min of processor time for the next move.
6.	. . .	O-O	
7.	e3	d6	
8.	Be2	a6	
9.	Na3		This knight will find it difficult to participate in the game.
9.	. . .	Nc6	Along these moves CHESS 4.6's evaluation function was telling it that Black stands the equivalent of a half pawn to the good and rising! This is based on its superior central influence and more active pieces. Note that the move 7. e3 has made White's QB very passive, though there was little choice.
10.	Nc2	Nf5	With an average of 4 min of processing time CHESS 4.6 is having little difficulty in predicting Browne's moves.
11.	O-O	Qd7	Now 12. Bd3 h4? 13. Nd5 Ng4?, 14. e4 was the predicted sequence, but it does not make much sense.

12.	b3		How else can White hope to find some life for his QB?
12.	...	Kh8	The first of a series of 'computer moves'. Though not bad in itself, the move indicates some difficulty in formulating a plan. Centralisation of the rooks with Rfd8 and Rac8 would be the way a human master would proceed to build up this position. Of course, this is the aspect of chess where computers show their greatest deficiency. Such moves also explain how the initiative can change hands so suddenly and frequently in computer games.
13.	Bd2	Rg8	
14.	Na4	Ba7	
15.	Bc3		Understandably Browne is trying to find some activity and at the same time trade off his passive N on c2. This is because Browne, as a GM, always tries to find the 'correct' move in a position. Had he better understood his opponent in this game, he would have made a waiting move (as David Levy frequently does against CHESS 4.6) when CHESS 4.6 might have weakened itself with . . . g5.
15.	...	h6	A useful move, providing a haven for the QB and taking care of back-rank mates.
16.	Rc1	Rad8	Now this pawn needed protection.
17.	Nb4	Nxb4	
18.	Bxb4		Now due to some technical problems CHESS 4.6 did not receive its full computation time and made a mild error:
18.	...	Qc7	18. . . . Qe6 with the idea . . . d5 would have been much better.
19.	Qe1		CHESS 4.6 still thinks itself a quarter pawn up, but with 19. Nc3 the game would have been roughly equal.
19.	...	Bc5	
20.	Bf3?!	Bd3	Now Browne starts to spend a long time at the board.
21.	Bxc5		The only move. Now if 21. . . . Bxf1?, 22. Bb6 wins a piece, but of course CHESS 4.6 does not make such oversights!

21.	...	dxc5	
22.	Be2	Bf5	
23.	f3	e4	
24.	f4		In the last few moves the game has changed somewhat in character. Nonetheless Black's control of the d-file gives him an edge.
24.	...	Bd7	
25.	Nc3	Qa5	A typical 'computer excursion' on the Q-side, though it is not bad here and Browne must think twice before initiating his K-side attack. Black, however, finds good defensive resources there and thus 26. Nd5!? deserves attention.
26.	Qh4		Played with an exaggerated flourish, yet Browne should have known that you cannot intimidate a computer program.
26.	...	Bc6	
27.	Rc2	b5	
28.	g4	b4	
29.	Nd1	Rd6!	Demonstrating the soundness of Black's last three moves. Now CHESS 4.6 expected: 30. g5 Nh7, 31. Nf2 Rgd8, 32. Bg4 Qb6, 33. Bf5 which is not best for either side.
30.	Nf2	Rgd8	
31.	Rd1	Rxd1+	
32.	Bxd1	Rd6!	At this point CHESS 4.6 had used 2 h 44 min of computation time to Browne's 22 min.
33.	Qg3		Perhaps best was 33. g5 Nh7, 34. gh gh (if 34. ... Rxh6, 35. Qe7 looks good for White), 35. Qe7 Qd8 when a tough endgame would follow.
33.	...	Qd8	Now CHESS 4.6 correctly predicts Browne's next 11 moves!
34.	Rc1	Rd2	
35.	g5	hg	
36.	fg	Nh7	
37.	g6		More solid was 37. h4, but then Black could safely snatch the QRP.
37.	...	fg	
38.	Qxg6	Qh4!	And now White is in great difficulty. Again Browne spent a long time deciding on his move. The end-game after 39. Qg3

Qxg3, 40. hxg3 would be hopeless for White.

39.	Qf5	Bd7

This move leads to a won ending for Black, though even stronger was 39. . . . Ng5! with threats of Rxf2 and Nf3+. White's checks would quickly run out after 40. Qf8+ Kh7, 41. Qf5+ g6. Then if 42. Qf4 Qxf2+!.

40.	Qf4	Qxf4
41.	exf4	e3

This wins a piece for two pawns. 41. . . . Nf6 was a more cautious way to retain good winning chances.

42.	Ne4	e2
43.	Bxe2	Rxe2
44.	Nxc5	Bc8

Computer programs usually lack the special-purpose knowledge required to win end-games such as this one. They do not know what transpositions to allow and do not have the refined positional judgement which is required. This move gets Black into a little trouble. Much more lethal-minded was the move 44. . . . Bh3!? with the ideas of mate. A continuation might be 45. Nxa6 Nf6, 46. Nxb4 Ng4, 47. Nd3 Rg2+, 48. Kh1 Rxh2+, 49. Kg1 Rg2+, 50. Kh1 Rg3 etc.

45.	Rd1	Re8
46.	a3!?	

Browne knows that he must trade as many pawns as possible.

46.	. . .	ba
47.	Ra1	g5?!

A dubious move because it offers to trade more pawns. Better was 47. . . . Rf8, 48. Nd3 Rd8, 49. Ne5 Rd2! Then if 50. Ng6+ Kg8, 51. Ne7+ Kf8, 52. Nxc8 a2 and Black wins. Also good was simply 47. . . . Re2.

48.	fg	Re5
49.	b4?!	

49. Nxa6 Bxa6, 50. Rxa3 would lead to the ending R + B + N *vs*. R which is a theoretical win, but here the White Q-side pawns might cause problems and it is questionable whether CHESS 4.6 would have the technique required in any case.

49.	. . .	a5!?

Again White could enter the above-mentioned ending, but under much less favourable circumstances.

50.	Nd3	Rxg5+
51.	Kf2	ab
52.	Nxb4	Ra5
53.	Ke3	

Here Browne misses a good drawing chance with 53. Ra2 (Not 53. Nc2 a2, 54. Nb4 Ra4!, 55. Nxa2 Be6 with Black winning). Then if 53. . . . Bf5, 54. Ke3 Bb1, 55. Ra1 a2, 56. Kd4 White should still draw.

53.	. . .	Be6
54.	Kd4	Ng5!
55.	Nc2	

This piece ends up playing a vital role. Browne offered a draw here which the Northwestern camp turned down in the 'interests of science'.

55.	. . .	a2
56.	Nb4	Ra4!
57.	Kc5	

Browne declines his last chance for R *vs.* R + B + N. He originally intended 57. Kc3 which would have met the same response.

57.	. . .	Ne4+!
58.	Kb5	Bd7+!
59.	Nc6	Nc3+!
60.	Kc5	Bxc6
61.	Kxc6	Rxc4+
62.	Kd6	Rd4+
63.	Ke5	Rd1

A pretty series of tactical moves.

and Browne resigned. An impressive finish by CHESS 4.6.

Key to special symbols used:
! — excellent move;
!? — enterprising but possibly unsound;
?! — interesting, but probably unsound;
? — a mistake.

Introductory Note to Chapters 8 and 9

Sixteen years downstream from publication of the first of these articles, the national grid conjectured in its last paragraph is at length coming into existence in the form of the General Post Office's *Prestel* system. As for *The Death of Paper* which follows it, its fantasy of paperless typists has now been fulfilled by today's reality of word processing. But the emergence of the desk-top computer-on-a-chip was not foreseen by this or any other author. Yet it could have been predicted from simple extrapolation of cost and size trends.

CHAPTER 8
Amplifying Intelligence by Machine

Artificial intelligence is a phrase which means only what it says. Artificial limbs, teeth or kidneys are mechanical contrivances which perform the functions of the corresponding biological structures. With artificial intelligence we aim to mimic the brain, or rather that small subset of the brain's population of neurones which is concerned with thought. The mechanical contrivance that most immediately suggests itself for the purpose is the electronic digital computer. This contrivance by itself will do nothing, so that the attempt to create artificial intelligence consists in writing computer programs of a special kind — in the first place geared to the manipulation of primarily non-numerical information and in the second place endowed with self-optimising properties that represent learning ability.

What progress has been achieved towards this goal, and where exactly do we stand today? The earliest criterion to be proposed for judging whether a machine could think was that of Turing, who suggested about 15 years ago that we should measure the machine's claim by its ability to fool us under cross-questioning into mistaking it for a human being. On this simple criterion, progress might be judged to be impressive.

Joe Weizenbaum's program, called DOCTOR, enables the machine to conduct a psychiatric interview, questioning the patient through the medium of a teleprinter, and responding in an apparently intelligent fashion to his replies. Sixty per cent of patients allowed to converse with the program refused to believe afterwards that they had not been connected to a flesh-and-blood doctor. 'No computer could possibly understand me that well!'

Unfortunately such an example, far from proving that the day of artificial intelligence has dawned, simply shows up a weakness in Turing's criterion. For if we go behind the scenes the illusion that this program is in some sense *understanding* what is said to it abruptly evaporates. The program in fact works according to the same simple bag of tricks that we ourselves use when we wish to give a companion — at a cocktail party, say — the illusion that we are following his remarks attentively, while we are really not listening at all. We pick up the odd key word or phrase (like responding to the key word 'mother' with 'Tell me more about your family') and if we are really stuck we just repeat our companion's last remark in an interrogative sort of way. The fact that the method works, and can actually be implemented by a fairly simple computer program, tells us more about human intelligence than it does about artificial intelligence.

I am not sure that anyone since Turing has proposed an intelligence test suitable to be applied to machines. Some American research workers are engaged in trying to construct computer programs which will do well at standard IQ tests and Marvin Minsky's group at MIT have some impressive achievements of this kind to their credit. In general, if we are told the range and nature of the tasks which a given program can perform, it would be useful to have rules for translating this information into an assessment of the intelligence of the performer.

First, unless a mental task is *difficult* its successful performance obviously does not count for very much. Second, even if the problem *is* difficult we do not call the solver intelligent if this particular task is the only thing he can do. There are many celebrated cases of *idiots savants*, that is, persons who are able to perform mental miracles within a single narrow category — such as calendar reckonings — yet are otherwise of subnormal intelligence. This is the present state of affairs with computers. A program for differential equations will not make any showing at all at draughts, or *vice versa*. The computer programs of today are all *idiots savants*. If the special task which a program performs is a very difficult one by human standards (as with differential equations or draughts) we may call the program 'powerful', but we still do not call it 'intelligent'. I shall return later to this question of adaptability to a wide range of problems and consider first the concept of 'difficulty' in more detail, for unless we can define and measure a problem's difficulty we cannot begin to talk scientifically about the amount of intelligence involved in solving it.

Subjective 'Difficulty'

The first step, I think, should be to categorise difficulty into subjective and objective elements. Subjective elements of difficulty are those which stem from the human peculiarities of the would-be solver, without in any way involving the intrinsic nature of the task. For example, the form of representation of the problem may greatly affect the solver's performance. One and the same problem may appear in the guise of a story about cannibals and missionaries, or about foxes and geese, or about x's and y's, and the solver may find it easier to get to mental grips with it according as he happens to be a colonial administrator, a farmer or an algebraist respectively.

A rather deeper example of *representation* is provided by the task of playing tic-tac-toe, otherwise known as noughts and crosses. The entire logic of this game, as was first, to my knowledge, pointed out by Frank George of Teaching Programmes Limited, Bristol, can equally well be represented as a purely numerical game, not involving at all the placing of marks in a 3 X 3 array. According to the numerical version, which Alick Elithorn of the Royal Free Hospital, London, tells me I should call 'number scrabble', play commences by laying nine numbered counters on the table, marked with the digits 1 to 9. Moves alternate, a player's move consisting in taking a counter from the table. The first player to have three counters in his possession which add up to 15 is the winner.

At first sight it is not in the least obvious that this game is logically equivalent to tic-tac-toe. The isomorphism follows from the fact that the tic-tac-toe board can be numbered with the digits from 1 to 9 to make a magic square (Figure 8.1) in which every row, column and diagonal adds up to 15.

The winning 'three in a line' pattern of tic-tac-toe thus corresponds in number scrabble to the possession of three digits adding to 15. The great difference from the human player's point of view is that although the two games are logically identical, number scrabble is far more difficult than tic-tac-toe. Two observations are relevant on this score. First, if one challenges someone at number scrabble without revealing that the game is really a recoding of tic-tac-toe, it is easy to win merely by preserving a mental image of the tic-tac-toe board throughout the play. Second, I have tried the experiment of allowing three colleagues who were ignorant of tic-tac-toe (the game is apparently not known on the Continent of Europe) to learn both the rule and the play by unassisted trial and error. They learned extremely quickly, whereas the equivalent learning task with number scrabble seems to be beyond most people. Evidently human experience of manipulating two-dimensional visual information here plays

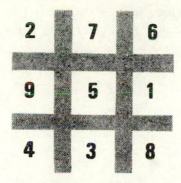

FIGURE 8.1.

an overwhelming role in deciding what is and what is not subjectively 'difficult'.

In the world of industrial technology, computer-aided design now offers exciting prospects of exploiting this specially human talent for 'seeing' a problem. By linking a skilled designer *via* a computer to visual display devices, we can allow him to build up trial solutions to, say, a network problem in forms which he can inspect and handle smoothly: he and the computer then converse by means of drawings. The same information expressed in the form of numerical co-ordinates, although logically the same problem, would be psychologically completely intractable, and a computer program, unless itself extremely 'intelligent', could then be more hindrance than help.

Objective 'Difficulty'

Now let us consider objective difficulty. To make things easier I will restrict attention to mental tasks for which solutions exist, and I will assume that these are attainable by performing a succession of steps leading from the starting state of the problem to the solution. I shall put forward the idea that the objective difficulty of a task is compounded of two quite separate and separable properties, each of which can be defined in terms of the solution strategy itself. The first property is the *information-content* of the strategy, and the second is its *complexity*.

Information-content is to be understood, in the spirit of the classical

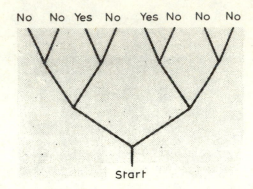

No   No Yes No   Yes No   No   No

Start

FIGURE 8.2.

definition, to mean the negative logarithm (usually expressed to base 2) of the probability that a purely random succession of choices would give the same solution path. Thus if the problem were to thread the three-level binary maze shown in Figure 8.2, we can easily see that the information-content of the successful strategy is just two bits (not three bits because the first of the three binary choices is indifferent; either 'left' or 'right' will do).

A particularly convenient illustration of the extension of this approach to non-trivial problems is a mental task which has been extensively used by Alick Elithorn for testing the intelligence of human subjects. Figure 8.4 shows a scaled-down example of the kind of maze he uses. What the subject has to do is to start from the bottom and to find a path to the top, passing through as many dots as possible. At each point in the lattice he has just two choices, left and right, since the path must always move upwards. In this example I have marked the solution strategy with arrows (note that there are some alternative paths). It can be expressed in a kind of logical notation as follows:

> L *and* L *and* L *and* ((L *and* R) *or* (R *and* L)) *and* R *and* ((L *and* L *and* (L *or* R)) *or* (R *and* ((L *and* R) *or* (R *and* (L *or* R)))))

The random probability of following this strategy can easily be computed by making the following substitutions in the above expression:

FIGURE 8.3.   *Early tools*. A step beyond unaided muscle power, a step behind powered tools and a considerable distance behind the exoskeletal harness in which motor impulses from nerves and muscles are picked up and fed to artificial muscles. Machine intelligence aims to provide means for amplifying brain power.

    for L read ½
    for R read ½
    for *and* read ✕
    for *or* read +.

Simple arithmetic then gives the answer 5/256. The information-content of best strategy in the example shown is thus $-\log_2 (5/256)$, or approximately 5½ bits. We would then say that in respect of the informational component of difficulty the problem of Figure 8.4 is less difficult than the specimen in Figure 8.5.

Here the best strategy can be written:

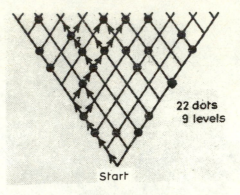

FIGURE 8.4.

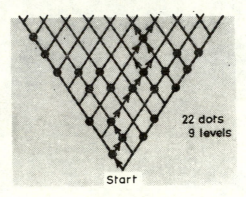

FIGURE 8.5.

L *and* R *and* L *and* R *and* R *and* R *and* ((L *and* R) *or* (R *and* L)) *and* (L *or* R)

with a corresponding information content of 7 bits. It is interesting, thought not surprising, that experimentally this measure does correlate quite well with the difficulty experienced by the human problem-solver.

Now let us turn to the second objective component of difficulty, the complexity of the strategy. This is very much more elusive and difficult to pin down into a rigorous numerical measure. But in cases as simple as the problem mazes under consideration we can do this in a crude and tentative fashion. Most of us, if asked for a *simple* rule of action would feel that we had to produce something which did not take us very long to describe. In the case of Elithorn's mazes, solution paths consisting of as few 'legs' as possible would score well from this point of view. Once again we would judge the second maze (Figure 8.5) the more difficult, this time on grounds of strategic complexity. There is good evidence from work by D.M. and M.G. Davies at Liverpool that this feature does contribute to the difficulty experienced by would-be solvers and hence that it is relevant to determining how much intelligence is involved in a given problem-solving activity.

I shall now leave Elithorn's very fascinating and instructive material, and give another illustration of components of difficulty, this time using an aptitude test under study in our group in Edinburgh.

We have been using a simplified version of the famous 'Fifteen puzzle'. In our version, the 'Eight puzzle', 8 numbered squares are free to be slid around in a 3 x 3 frame, there being always one space empty. The goal position is defined arbitrarily as:

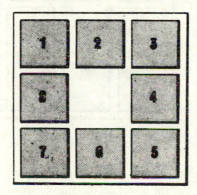

FIGURE 8.6.

From some starting position the object is to rearrange the pieces into the goal configuration by means of as short a sequence of sliding moves as possible. The three components of difficulty instanced above all re-appear in our puzzle as follows:

1) *Representation.* The outward form of the puzzle can be recoded by repainting the square pieces, so that, for example, the goal position

FIGURE 8.7.

FIGURE 8.8.

looks like Figure 8.7 or 8.8 above. The logical structure of the problem remains the same, but the effect on difficulty can be startling.

2) *Information-content*. The number of bits of information required to specify a minimal path to the goal is roughly proportional to the length of the path. It is found that the solution paths found by human solvers also run in proportion over almost the full range of minimal path lengths, the typical ratio for untrained subjects being three times minimal.

3) *Complexity of strategy*. A very simple, but moderately effective, strategy relies on building up a chain of sequentially numbered pieces, in cyclic order round the frame. Starting configurations which already

possess two or three pieces in sequence are solved more efficiently than configurations, no more difficult in other respects, which do not possess clues prompting to a simple strategy.

On the criterion of difficulty, advanced computer programs have already far outstripped the great majority of mankind and are beginning to peg level with the performance of highly trained minds — whether in game-playing, traffic scheduling, theorem proving, engineering design, and so forth. So much for difficulty of task. On my other criterion of intelligence, that of generality of application, it is a different story.

The Search for Generality

The attempt to write a program to simulate the human thinker's capacity to be a Jack-of-all-trades, even though a master of none, is a daunting business. Newell and Simon in the United States have made great pioneer progress, but their program, hopefully christened the General Problem Solver, seems to have run about as far along its particular track as it is capable of getting. Their approach is based on the formulation of a generalised problem as a set of *states* together with a set of transformations which may be applied to a state to turn it into another state. Given a means of measuring the difference between any intermediate state and the desired state, or goal, we seem to have all that is required for a purely mechanical strategy for applying transformations successively until a solution is reached. The only further requirement is a way of mapping any particular problem, such as the propositional calculus, for example, or chess, on to this platonic ideal of a problem.

The way in which formal equivalence can be set up between apparently totally unrelated systems is illustrated in Figure 8.9 where two simple problems, the Eight-puzzle and a problem of algebraic manipulation are both mapped on to a single formal representation — essentially the one proposed by Newell and Simon.

My own research group now has a working program thanks to Mr J.E. Doran, capable of tackling in a moderately creditable fashion any problem which can be fitted into this same framework of states and transformations. We set store upon the future growth of this promising infant, for it has already acquitted itself adequately on a well-known, and recalcitrant, problem of commercial importance — the 'Travelling Salesman' problem.

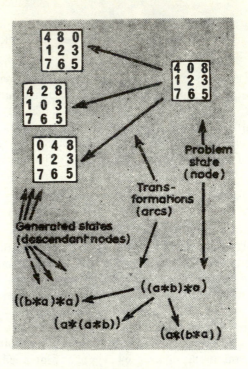

FIGURE 8.9.

But we have to admit that as yet it has scarcely come to grips with one of the most vital requirements for any general problem-solving program, namely a means of communicating to the program in a flexible and easy fashion descriptions of the problems to be solved other than by *ad hoc* programming. I myself believe that a necessary condition of success is the availability of powerful 'on-line' computing systems which will make it possible for the program to put questions to the user. As yet there are no such systems in Britain, although there are a number of groups who are now quite busy trying to bring them into being. When these exist and have developed into a national grid, equipped with easy-to-use information-retrieval and problem-solving programs, then we shall truly be able to say that the fourth stage of the computer revolution has dawned and that the concept of 'amplified man' in his intellectual aspect may become an engineering fact.

CHAPTER 9
The Death of Paper

Most of us view the headlong rush of computers into our lives with apprehension, as an invasion by a race of conquering aliens. In reality we should think of ourselves as a beleaguered garrison which at the eleventh hour sees on the horizon the dust of the relieving column.

The role of computers as increasingly powerful ancillary troops in scientific and technological advance speaks for itself. In this article I shall speak of a peril which now faces the more advanced and civilised sectors of the globe, threatening the mind and spirit of man rather than his body — the rising tide of paper.

There was a time when only those foolish or unfortunate enough to do clerical work in business and administration were subjected to the paper menace. The scholar or scientist could still keep paper at bay, consulting books when he wished and generating written texts only at his own volition. Now, not only the scientist but every housewife, taxi-driver and shopkeeper is deluged with forms, certificates, notices, licences, tickets, cheques, stamps, permits, claims, memos, and the like, not to mention the flood of newspapers and magazines which enter his door. Little wonder that industrial man has lost himself, and cannot tune his mind to the messages of literature, art, science or history when his anxious fingers riffle eternally through a paper torrent.

But are not computers themselves among the most active paper-producing agencies devised by man? On the face of it, certainly so. The CDC 6600 computer at CERN, the great multi-national high-energy physics research centre in Switzerland, operates 3 high-speed printers simultaneously, *each* capable of producing 1 cwt of paper a day. Fortunately, however, my task here is to describe the computers of the future, not of the present, and these future computers (as I shall explain) will be as sparing of paper as their predecessors have been prodigal. More than that: they will make totally unnecessary the production of paper at all, whether by themselves or by anyone else.

The new computers, of which the prototypes are already with us, keep this potentially voluminous output internally, only displaying on demand, and in transient form, the particular items which the user wishes at that moment to see. The display is on a screen like a television screen, and the user can change, delete, or add items on the display before dismissing the picture and returning it to the internal store. At no stage is paper involved. If I want to see my birth certificate, or a page of mathematical tables, or a reproduction of a Picasso drawing, then, provided these are among the items filed in the computer memory, I can call them up on the screen, inspect them for as long as I like, and be done with them until I next wish to call them. Human beings use paper because of the frailty of human memories. The gargantuan growth of computer memories will change all this. The CDC 6600 machine at CERN has today a total memory capacity equal to the content of an entire library of books. Once these vast memories, with appropriate search and retrieval routines, are harnessed to the human user, it is the death of paper.

To understand how this can come about we must envisage a world, perhaps 15 to 20 years in the future, in which the telephone network and the thousands of variously owned and managed computer installations have fused into a single integrated communication-computation network or 'grid'. In that world, computing consoles will abound as plentifully as telephone sets today, all connected to the National Computational Grid. Every typewriter in every office will type, not on to paper, but directly, via a telephone line, into the grid, from which the page image will be simultaneously displayed on a screen in front of the typist. Each console will have a telephone, not only for person-to-person speech, but also for person-to-computer and computer-to-person conversation — for by that time the problems of speech recognition, at least for a kind of 'Basic English', will have been solved. The console will have, in addition to visual displays, one or more visual input devices (computer 'eyes'). The apparatus of discourse will thus be fully comparable with that at present enjoyed by one human when communicating with another; he can speak, listen. draw diagrams, type a note, be shown a document and in turn present a document to view. There will be a further feature: the visual display will produce, on demand, a photographic 'hard copy' of the image currently displayed, supposing that on some occasion he will actually want a piece of paper to carry away with him.

One of the first kinds of paper to become obsolete will be money.

When every cash till is a terminal to the grid, the marking up of a purchase will simply debit automatically the customer's centrally stored bank account.

It may well be that as our citizen of tomorrow peers moodily at his morning tabloid projected on to the breakfast table top under dial control, or browses through his electronically stored books with his console display, or has his households accounts done for him on the household type-writer, or follows through the steps of a teach-yourself Egyptology 'correspondence course', enjoying a two-way communication by voice and diagram with his electronic home tutor, or pursues at will any of the infinitely branching paths in the labyrinth of art, literature, science and mathematics which will be at his instant service — it may well be that the only need he will have for paper will be his identity card carrying his code number. Paper will then at last have come into its own for its true and natural purpose, an unsurpassed material for *disposable* goods — clothing, crockery and the more private needs of the household.

Introductory Note to Chapters 10 and 11

Computer-controlled robots are widely familiar today. Some of the very first receive mention in the following two articles. Where are they now? Although neither SHAKEY nor FREDDY assayed the San Diego trail proposed by McCarthy, their achievements were various, continuing long after these articles were written.

Old robots never die, but already the hungry generations tread them down. A decade after the birth of these academic devices, Unimation Corporation alone has installed more than a thousand for factory work, and computer-controlled robots figure in expansion plans for space technology and sea-bed exploration.

CHAPTER 10
Clever or Intelligent?

It is fashionable to talk of the computer revolution as the second industrial revolution. The first revolution mechanised muscle power, and today motor cars, ships and aeroplanes, power stations and factories sustain a population which has multiplied fivefold since the lifetime of James Watt. But now we see the beginning of the mechanisation of brain power. It is even possible that just as machines have outstripped muscles, so computers may some day outstrip brains.

The fact that we are still in the first onrush of the second revolution tends to blind us to the pace and acceleration of change: even now a third revolution is gathering momentum – the harnessing of mechanised brain power to mechanised muscle power – i.e. intelligent robots. I am not going to speculate about the uses to which these new and rather awesome creatures will be put. The early guesses in the case of horseless carriages or heavier-than-air machines seem half-baked when we look back. I am going instead to talk about the work of a few laboratories scattered around the world, including our own in Edinburgh, where scientists are trying to develop a design philosophy for automated intelligent behaviour. Just as at earlier stages in technological history one would find groups developing the hovercraft philosophy, the jet philosophy, the flying-machine philosophy or the philosophy of the horseless carriage, so now a number of American laboratories, and three in Britain – at Edinburgh, Aberdeen and Sussex Universities – are trying to develop machine intelligence.

In fields like space exploration, which require billions of dollars of expensive equipment, American supremacy is perhaps inevitable. But machine intelligence is open for rapid development with a research budget of relatively modest size. It may therefore be of interest to talk about what research workers in this field are actually trying to do and what are the problems of programming a robot to act intelligently.

It was an Englishman, Alan Turing, who first suggested in 1953 that board games like chess or draughts could provide models through which

150

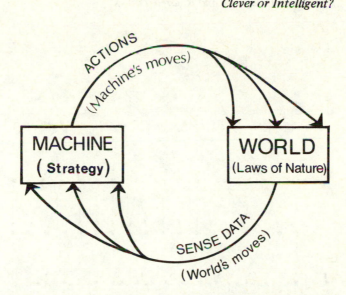

FIGURE 10.1.   The interaction between the machine and its environment depicted as a game against Nature. Equipping the machine with strategies for this game is the subject matter of machine intelligence.

the design principles for intelligent robots could be investigated. The basic idea is shown in Figure 10.1. Here the interaction between the machine and its environment is depicted as a game against Nature. One player's strategy (Nature's) is fixed. We can't do anything to change it. What about the other player's strategy? Inventing strategies for the machine side of the game is precisely the research worker's craft, just as inventing stories is the novelist's craft, or inventing music is the composer's. Figure 10.2 shows the general structure of a game. In such a tree diagram the blobs represent states of the board and the lines represent legal moves. To put this in terms of the game against Nature, Figure 10.3 shows a fragment of a decision tree describing an imaginary problem from real life. Each node on the diagram represents a possible state of affairs and the lines represent legal moves. To put this in terms of the game against Nature, Figure 10.3 shows a fragment of a decision tree describing an imaginary possible state of affairs and the lines radiating from each node indicate possible reactions of the world, or possible lines of action for the intelligent planner, according to whose turn it is to play.

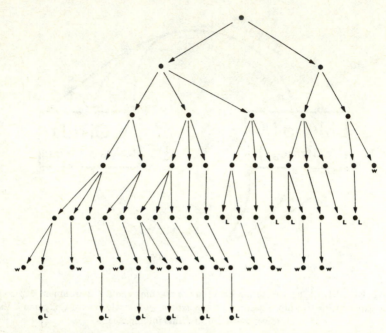

FIGURE 10.2.   A game tree. Here the blobs represent states of the board and the lines represent legal moves. The terminal nodes have been labelled with W for a won game (for the 1st player) and L for a lost game.

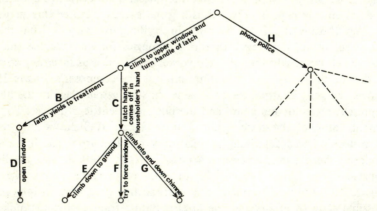

FIGURE 10.3.   Part of a planning tree for a householder locked out of his house. Each node represents a state of affairs and the lines represent possible courses of action or reaction on the part of the household or the environment, according to whose turn it is to play.

Type of strategy	Type of behaviour
A only	reflex
A + B	adaptive
A + B + C	simple cognitive
A + B + C + D	complex cognitive

<u>Key</u>

"Clever but mindless"
{
A: fixed stimulus–response tables
B: modifiable stimulus-response tables
}

"Intelligent"
{
C: cognitive maps using lookahead
D: inductive reasoning and planning
}

FIGURE 10.4.   Diagram showing two types of strategy: 'clever but mindless' and 'intelligent' together with the types of behaviour they give rise to.

As indicated in Figure 10.4 strategies come in two sizes: (1) clever but mindless and (2) intelligent. An example of category A in the realms of games would be a Noughts and Crosses strategy in the form of a look-up table: that is, a table showing a recommended move for every possible board position. In an adaptive version, table entries would be subject to change in the light of accumulating results of past plays. A working example is a machine which I made out of matchboxes (Michie, 1963) many years ago to illustrate the principle. Each box is in effect an item in a look-up table, corresponding to a particular board state. In each box are beads of different colours, corresponding to different moves. By a reward-and-punishment regime involving adding or subtracting beads for those used in a given play, elementary trial-and-error learning can be demonstrated. It turns out that even this elementary kind of learning results, on a purely automatic basis, in surprisingly complex forms of behaviour. Figure 10.5 shows the performance of a computer program constructed roughly on the match-box principle (see Michie, 1968) learning to master a quite difficult motor task, which it had never seen before, namely balancing a pole on a motor-driven cart. This task was originally formulated, in a different context, by Donaldson in 1960. Here the pole-and-cart

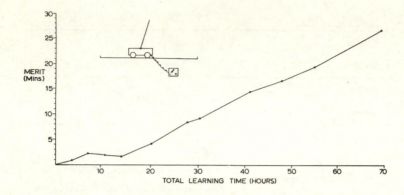

FIGURE 10.5.   A 'clever but mindless' strategy which shows adaptive behaviour using a simulated pole and cart system harnessed to a learning computer. The curve shows average time-until-crash plotted against learning time. Thus after 70 hours of pure trial-and-error learning, the pole is balanced for approximately 30 minutes.

system is simulated in a second computer connected to the learning computer by a high-speed data link. Every 15th of a second the pole and cart system sends a signal across the link, and the learning program must immediately reply either 0 or 1. The reply imparts either a leftward or rightward drive to the cart's motor. No interpretation of the state signals is given to the learning program, nor any information as to the significance of the 0 (left) versus 1 (right) choice. But as can be seen it manages to piece together an adequate control strategy.

In the robot realm an early triumph in this category was Grey Walter's electronic tortoises which were capable of quite involved goal-seeking behaviour. Latter-day descendants include a wide variety of automatic control and guidance systems used in space navigation, ballistic weaponry and the like. Why do we call these clever machines mindless? What is the equivalent of a 'mind' when furnishing a robot? The key feature is the ability to construct and store internal models of the world from the fragmentary incoming stream of sense data, and to use these as the basis of planning. We owe this formulation to Craik (1943) the gifted experimental psychologist whose early death in 1945 robbed Britain of one of her brightest scientific hopes.

An example of an internal model is the 'cognitive map' which we construct when we learn to find our way through a maze, or even our own house. A table of situation-action pairs will get us just so far until something goes wrong — like finding ourselves locked out. It is then crucially important to us that *in our model,* as in outside reality, the chimney connects the roof to the inside of the house. This is what makes it possible for us to include for example the move 'climb into and down the chimney' in the planning tree of possibilities shown in Figure 10.3.

Growing planning trees inside the computer is something we now understand moderately well how to do. Teaching the machine to construct useful cognitive maps from fragmentary sense data is less easy, but a beginning is being made in the various laboratories. At Stanford University the Artificial Intelligence laboratory is developing a computer program to co-ordinate one sensor (TV camera) and one effector (mechanical hand) to perform a fairly stereotyped task the sort that a two-year-old child can manage (piling bricks). Nonetheless, the program must perform elaborate feats of model construction and verification as the basis for planing each move. This is even more true when we come to integrated behaviour, as planned by Stanford Research Institute, whose robot is shown in the accompanying photograph (Figure 10.6). Figure 10.7 shows a local essay on a related theme in the form of Freddy, the newly constructed 'real world interface' for our research computer. Designed and built in Edinburgh by Steve Salter, it has now been successfully connected to the computer by Harry Barrow. We talk to it through a teleprinter terminal in the programming language POP-2, developed by Rod Burstall and Robin Popplestone. The kind of task which we are attempting to teach it is to find and identify simple objects in its world, to move them so as to satisfy stated conditions, and to make and print out maps of its world from time to time.

In Figure 10.4, I distinguished lower and higher forms of intelligent behaviour. The kind of planning described so far is still at the lower level, a level extensively explored in research on computer programs for games such as chess. Here a plan takes the form of a chain of moves or at most a branching sequence. But there is a higher level of problem-solving in which this kind of planning is simply not good enough. At this higher level, planning actually takes the form of generating new programs, later to be executed when the moment for action arrives. The idea of a computer program clever enough to write and execute its own programs has flitted across the machine intelligence scene since early days. We are now, I think, beginning to see how to handle what was always an exciting

FIGURE 10.6.   The robot developed by the Stanford Research Institute (photo from Nilsson *et al.*, 1968).

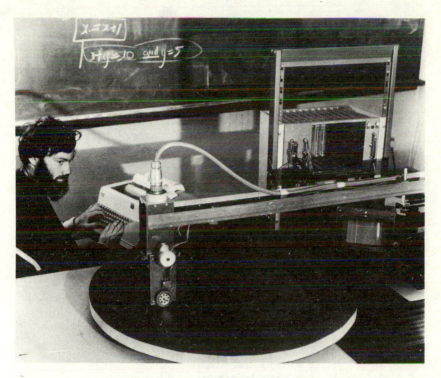

FIGURE 10.7. Freddy-1, the first Edinburgh robot, searching its world for objects. Robin Popplestone, one of the authors of the programming language POP-2, is instructing it via the teleprinter.

concept but is at last becoming a practical one. Imagine a robot equipped with wheels, steering wheel, etc., in other words a cognitive automobile. Suppose it is now hundreds of miles from San Diego. John McCarthy has suggested the following problem, which might be put to our robot:

1) At each filling station a map is obtainable on request;

2) Each map has an arrow pointing towards San Diego;

3) Each quadrant of each map contains at least one filling station;

4) Kindly produce and execute a plan for planning and getting to San Diego.

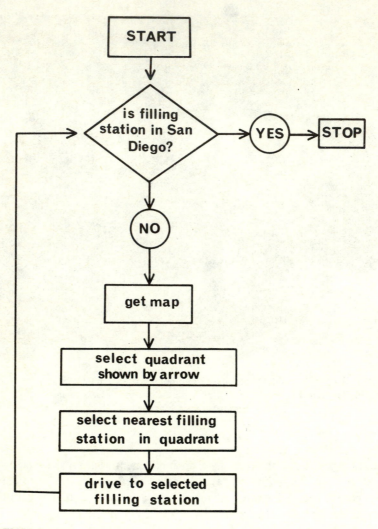

FIGURE 10.8.   Answer to the San Diego problem, shown as a flow diagram.

Figure 10.8 shows the winning answer. Of course this plan is just an outline scheme and in practice some of the constituent boxes themselves pose problems for which miniature plans of the same form would have to be generated, e.g. for selecting the nearest filling station in a quadrant of the map, let alone for handling the actual driving.

We do not as yet fully understand how to program a computer so that it will generate high-level plans of this type. But thanks to some rather abstract work on methods of proving things mathematically about computer programs, lines of approach are beginning to come into view. In the United Kingdom active schools both of machine intelligence research and of abstract programming theory have been rapidly gaining momentum and through the annual series of Machine Intelligence Workshops which we hold in Edinburgh, and through many other media these schools are keeping in lively touch with each other. With a little luck we may see the new insights being applied during the next few years to endow experimental 'real-world' devices with powers of learning and planning which may seem surprising even to their inventors.

Although we cannot say now in 1969 precisely how advances in machine intelligence will transform our society and the lives of our children we know that the transformation will be far reaching, whether through the automatic control of factories, or intelligent machines for operation in remote environments like space or the ocean bed, or the arrival of the conversational computing terminal in the home as an intelligent extension of the domestic television set. There is a chance for Europe to move into the lead in this exciting area of new technology, the foundations of which were in fact laid in Europe some thirty years ago.

REFERENCES

Barrow, H.G. and Salter, S. (1969). Design of low-cost equipment for cognitive robot research. *Machine Intelligence 5* (eds. B. Meltzer and D. Michie). Edinburgh: Edinburgh University Press.

Craik, K. (1943). *The Nature of Explanation.* Cambridge: Cambridge University Press.

Donaldson, P.E.K. (1960), Error Decorrelation: a technique for matching a class of functions. Proc. III International Conf. on Medical Electronics, pp. 173–178.

McCarthy, J. Personal communication.

Michie, D. (1963). Experiments on the mechanisation of game learning. Part 1. Characterisation of the model and its parameters. *Computer Journal* 6, pp. 232–236.

Michie, D. (1961). Trial and Error. *Science Survey*. Part 2. pp. 129–145. London: Penguin.

Michie, D. (1968). 'Memo functions' and machine learning. *Nature*, 218. pp. 19–22.

Nilsson, N.J., Rosen, C.A., Raphael, B., Forsen, G., Chaitin, L. and Wahlstrom, S. (1968). Application of intelligent automata to reconnaisance: Final report. Prepared for *Rome Air Development Center*, Griffith Airforce Base, New York 13440. SRI project 5953. Menlo Park, California, Stanford Research Institute.

Turing, A.M. (1953). Digital computers applied to games. *Faster than Thought* (ed. B.V. Bowden). London: Pitman.

CHAPTER 11
The Intelligent Machine

Simple tasks which we perform a thousand times a day without thinking are in fact of profound intrinsic difficulty. The logical analysis of such tasks may soon be used to design a primitive intelligent robot.

Advances in computers have already taken us past the point where we ask: 'Can a computer beat the human brain?' To be taken seriously, the question must be qualified: 'Can a computer beat the brain at what?' It is a commonplace that computers can now beat us at arithmetic, invoice-handling, stock-market forecasting, project scheduling, airline booking, traffic control — even at chess a computer program has been developed for the big CDC 6600 machine which can give an international chess master a run for his money. So the centre of interest shifts to what computers cannot yet be programmed to do. To bring into focus in the reader's mind just where this unknown territory lies, I reproduce here some standard tests for two- to three-year-old children selected from the Stanford-Binet range.

The first is intended for children about two and a half years old. Five objects — a brick, a button, a toy dog, a box and a pair of scissors — are laid out in a row on the table. The child's response to commands like 'Give me the dog', 'Put the scissors beside the brick' is assessed. The second test, sorting buttons, is intended for children about three and a half years old. The equipment required is 20 buttons, 10 black and 10 white, and a box. The buttons are emptied from the box on to the table in front of the child, and the box cover placed beside the box ready for sorting the buttons. The adult conducting the test takes a button of each colour from the pile, saying 'See, the black buttons go in this box, and the white buttons go in that box. Now you put all the black buttons in that box and all the white buttons in this box.'

This trite looking material is presented in order to jolt the reader out of the preconception that such tasks are intellectually easy. They are easy, of course, if you already know the answers. Therefore they are easy

161

for adult human brains, which spend their every waking hour in a world filled with such tasks. So it cannot be stated too often that they are of profound intrinsic difficulty, as is immediately discovered by anyone who attempts to simulate such skills by computer — more difficult than every single one of the tasks listed earlier, more difficult than arithmetic, invoice handling, stock-market forecasting, project scheduling, airline booking, traffic control, more difficult than chess.

Just where is the difficulty in designing a computer based integrated cognitive system (ICS)? If we could answer this question in detail we would already be on the way to programming a computer to compete intellectually with human infants. (We dismiss the idea of pre-programming the computer for each individual test: the ICS like the infant, must be allowed no advance knowledge of what the examiners have in store for it.) But one or two points can be made, and rough headings mapped out.

The first point is that the obstacles to progress in machine intelligence research are no longer attributable to limitations in the speed or storage capacities of contemporary computers. The table on page 168 compares computer and brain in terms of raw information handling power. Further, the world's computing power is increasing by at least a factor of 100 every decade, so we can be fairly sure that such limitations will never again be a bottleneck. The difficulty lies not in shortage of computing power, but of the needed insights into mathematical logic and programming theory to enable us fully to harness it.

The next point is obvious, once stated. These simple tasks of recognition and manipulation seem to us trivial and unintellectual only because of our extreme familiarity with them. Whatever processes of mental discovery were involved in our early life have long since been pushed out of sight and our acquired problem solving skills have been converted into automatic behaviour, operating below the threshold of consciousness. We do not ask ourselves how we know that these objects are 'buttons' or what exactly we understand by the notion 'inside', or the action 'put'. We may once have done, but all this is forgotten. So in approaching the mechanisation of these skills we must somehow wash our minds clean of associations, and approach the analysis of such tasks as though we were re-born as disembodied spirits of pure thought. Naturally we cannot ever quite do this, and our special knowledge of real world problems continually obtrudes and distracts. But at least we can be on our guard.

Let us now look at the skills involved and see how we might break them down into parts. I shall consider two major subdivisions — one being to do with interpreting the inputs to any ICS to which I shall apply the general term 'recognition' and the other to do with specifying the outputs, which I shall refer to as 'planning'. Both activities will be found to lead on to processes of inductive generalisation.

To deal with real world objects, an ICS must be able to recognise them — not only to attach a name, but to associate with the name a cluster of properties and relevant facts. Recognition at this level must employ techniques of 'pattern-recognition' but cannot be achieved by such techniques alone, any more than machine translation can be achieved by grammatical analysis alone.

A great deal of work on the incorporation of intelligence into computer 'eyes' has been in progress at a number of American centres, notably in Marvin Minsky's laboratory at Massachusetts Institute of Technology and John McCarthy's at Stanford University; and also at the robot project at Stanford Research Institute. In particular I should mention Adolfo Guzman's work at MIT on the analysis of scenes composed of three-dimensional Euclidean objects (cubes, pyramids, wedges, etc.), work by J.A. Feldman, M. Heuckel, R. Paul and G. Falk at Stanford University, and the approach developed by C.L. Fennema and C.R. Brice at SRI known as 'region analysis'. All but the last have concentrated on the extraction of line drawings from the images formed by the TV camera, and/or the subsequent processing of line drawings. Thus they have been firmly based on techniques such as edge following and the extraction of such features as corners and curves. There is a reflection here of the decision to restrict analysis to regular objects. An authoritative review of picture processing from this standpoint has recently been published by A. Rosenfeld.

At Edinburgh we are concerned that the final system should not need to rely on Euclidean assumptions and should be able to cope with irregular objects such as cups, hammers, tobacco pipes, spectacles or Teddy bears. In consequence we have adopted the SRI 'regional analysis' approach, and have added to it a suite of procedures for forming and matching abstract descriptions. These procedures are derived from concepts found in a branch of mathematics known as graph theory. Our visual recognition system, developed by H.G. Barrow and R.J. Popplestone, is now able to inspect individual objects through the TV camera and identify them on the basis of stored abstract descriptions supplied by a human 'tutor'. Let us go through its operations, stage by stage. We will suppose that the camera is pointing at the tea-cup (see next three pages).

1) Input from the TV camera is turned into a video signal and held in computer store as a 64 x 64 array. Each cell of the array contains a number encoding one out of 16 possible levels of brightness. Experience indicates that further important gains in performance will be achieved by going up to a 128 x 128 array with 32 brightness levels.

2) In the spirit of 'region analysis' an attempt is then made to identify, not lines and edges in the visual image, but elementary regions composed of neighbouring points of similar brightness. Adjacent regions are then merged to combine separate regions which are 'really' a single region — that is, correspond to a single plane or other homogenous sub-object. For example merging will occur across a boundary if the average contrast across it is less than a pre-set threshold. The process continues until no more merging can be done, and we are left with a set which hopefully corresponds to the surfaces, holes, and other features of the object.

3) A description is now synthesised in terms of topological and geometrical properties of the regions and their interrelations. Thus 'region A is compact (roughly circular)', 'region B is compact', 'region A is adjacent to region B' 'C is inside B' 'the A/B boundary is convex with respect to A' and so on.

4) The description is now matched against a set of stored descriptions, and the best match yields an identification of the object together with a measure of confidence.

Quite a small set of properties and relations can be sufficient for identifying a range of everyday objects, each analysable into a small number of regions. This is just as well from the experimenters' point of view, as our present system takes 5 to 10 minutes over a single identification. A speed-up of a hundred times is needed. This should be attainable by further improvements in the program, by delegating some of the visual processing to a small fast satellite computer, and, eventually by transfer of the work to a more powerful 'main frame' than our present ICL 4130 machine. At present the abstract descriptions are supplied by the human user. A true ICS must be able to synthesise its own descriptions where necessary. To

(A)

(B)

FIGURE 11.1 (caption on page 167)

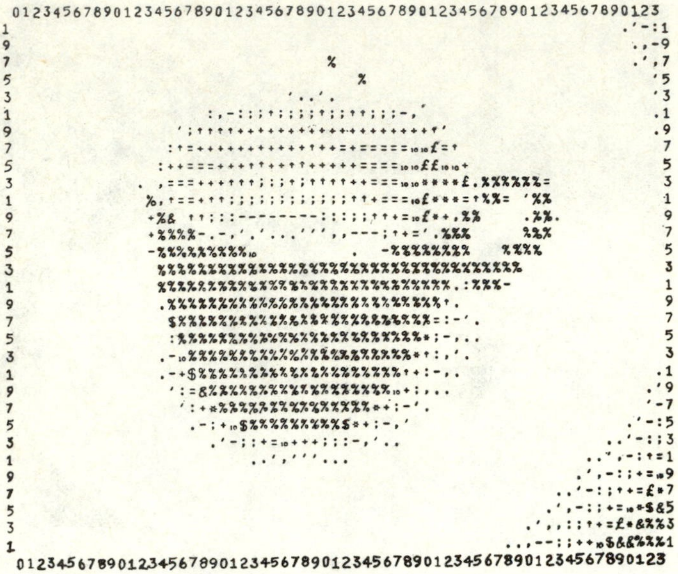

(C)

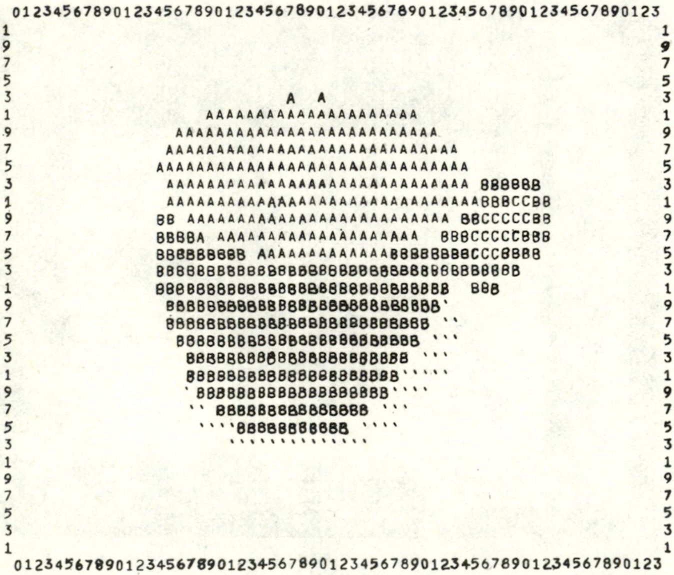

(D)

FIGURE 11.1 (caption on page 167)

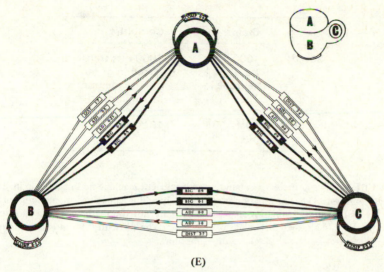

(E)

FIGURE 11.1.   EDINBURGH MARK I robot device (A) consists of a television camera pointing at an object, in this case a teacup. The viewing platform is mounted on ballbearings, so that operation of the wheels by the computer moves the platform around, changing the camera's view of the scene. The striped bumpers fore and aft signal contact, left and right bumps being discriminated. The robot uses a time-sharing system on the department's ICL 4130 computer. The TV picture of the teacup is progressively analysed by the computer system. The first picture (B) shows the tea cup displayed normally on a monitor screen. This picture is turned into a video signal and stored in the computer as a 64 by 64 array with 16 possible brightness levels. In the illustration (C) different line printer symbols have been used to show the 16 different levels. The next stage (D) is to identify regions in the image composed of neighbouring points of similar brightness. The hole in the cup handle, for instance, is represented by the letter C, and shadowing round the lower half of the cup by the symbol "'. The final stage (E) is to describe the regions in terms of their properties and relations. The numbers associated with the arcs indicate the strength with which the property or relation is present. Thus COMP (compactness) is the ratio of the area to the square of the length of the perimeter, normalized by the factor 4π. ADJ (adjacency) is calculated as the proportion of the perimeter which is in contact with the given adjacent region. BIG refers to the size of a region and DIST to the distance between two regions. Only a selection is shown of the many properties and relations handled.

	Brain	Computer
Speed:	1000 bits traverses 1 neurone in 1 sec.	1000 bits transferred in or out of core memory in 1 microsec.
Store:	$10^{12}-10^{15}$ bits	10^{12} bits

FIGURE 11.2.   Difficulties in building an intelligent machine no longer lie in inadequate hardware. The information storage and handling capacities of modern computers and the human brain (compared above) are not too dissimilar, even allowing for the brain's ultra-parallel mode of operation. The difficulty lies in programming the computer.

do this, powers of generalisation will be required, which takes us into a topic which we will raise again later.

Another problem we have tackled is concerned with planning. A plan can be defined as a sequence of actions, linked if necessary with further sensory tests, to convert the present situation into a desired situation. The ICS is supplied with (1) a description by the user of the desired state of affairs (2) a description of the present state of affairs constructed from sensory inputs and (3) a repertoire of actions.

The difference between simple recognition tasks and planning is summed up in the difference between the two questions: 'What is this?' (a cup is placed in front of the TV camera) and 'Find a cup' (a jumble of objects including a cup is placed in front of the camera). The first question is one of recognition only; the second requires recognition, but it also requires the synthesis of a complex and subtle sequence of actions and tests. It might, for example, involve spreading the objects out on the viewing platform in mutual isolation, inspecting each in turn, and stopping when a cup is identified. Each of these involves the formation, execution and revision of plans at a more detailed level. To spread the objects out, movements of a 'hand' must be devised and implemented, with continual adjustment of its relative position and the degree of separation of the objects.

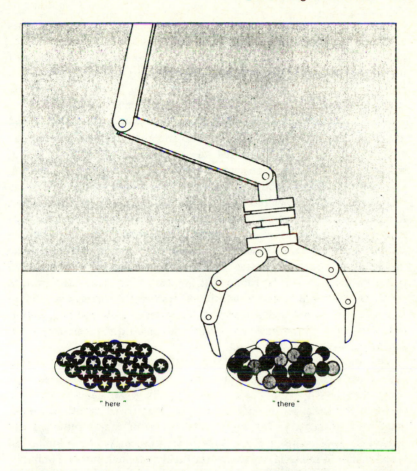

FIGURE 11.3. Ultra-simple world designed by Robin Popplestone contains only two places, 'here' and 'there'. In addition it has a blind 'hand' and a number of star-patterned objects. The task is to devise a program for planning how to achieve a state of affairs in which at least one star-patterned object is 'there'. Despite the triviality of the problem, writing the program is difficult for a machine.

A key phrase in the earlier definition is 'linked if necessary with further sensory tests.' If this phrase is removed, then we are dealing with simple planning. By this we mean the formation of a plan as a fixed sequence of actions — 'turn right, turn right, straight on, turn left,' and so on.

Complex planning on the other hand, performs tests on abstracted features of the task environment: 'turn right *when* you come to a roundabout, straight on *until* you see a church' and so on. Computer construction of simple plans is now well understood, and techniques of heuristic search have been developed to handle such problems by 'lookahead' processes from which as much as possible of the blind search is removed.

Complex planning, on the other hand, is beyond such methods, and techniques from the armoury of automatic theorem proving must be brought into play. The essential ideas were first formulated by J. McCarthy, and recently brought into sharp focus by work of C.C. Green at Stanford Research Institute. Consider the classic monkey-and-banana problem: a monkey is placed in a room which contains a pole and a box. A bunch of bananas hangs from the ceiling, well beyond the monkey's reach. Can the monkey get the bananas? Let us pose this question as a conjectured theorem, which is to be proved from a set of axioms defining the problem, and which states that there exists a possible situation in which the monkey has the bananas. It is possible, as Green has shown, to set up an automatic procedure which will prove the theorem, and produce as a side-effect the actual sequence of operations (go to the box, push it until it is under the bananas, go to the pole, pick up the pole, etc) which constitutes the needed plan of action.

The drawback of the approach lies in the cumbersomeness of present day procedures for operating on logic statements, and the speed with which they tend to ramify into a forest of irrelevant deductions, particularly if the initial set of axioms is of more than trivial size. Certainly for many very elementary planning situations, or for simple sub-plans embedded in more complex plans, direct search processes of the lookahead type familiar in game-playing can be profitably adopted. It seems unlikely, however, that use of the full apparatus of mechanised logic can be dodged for problems described in terms of general facts (there is a box somewhere in the room; the pole is at least four feet long; the monkey is strong enough to lift the pole; and so on). It is this lack of precise specification, common in real life problems, which prevents one reaching a solution by running in the computer a kind of cinematograph representation of sequences of events along alternative future action-paths — 'planning by simulation' — and forces an approach through formal deductive reasoning. This point is essential to grasp if one is to understand the kinds of problems which research in machine intelligence is now facing. Partly for this reason and partly to drive home the logical complexity even of feats of planning which seem trivial, I shall now consider an example using the simplest meaningful problem-universe which my colleague Robin Popplestone has found it possible to devise.

Popplestone's world contained two places only, called 'here' and 'there'. Actually even their twoness was not specified, so that 'here' and 'there' could be one and the same place: but this is not of importance. There are indefinitely many 'things' in this imaginary world, and any number of them may be located at one time in a given place. The world is inhabited by a robot, consisting only of a hand which is capable of three actions: 'goto' which moves the hand to a stated place (or, if the hand is already at that place, leaves it where it is), 'pickup' which transfers into the hand a thing chosen at random from the place where the hand is, and 'letgo' which has the obvious meaning. Note that the hand is anaesthetic and blind; it has no means of telling directly whether it is holding something, and no means of inspecting a thing held for the presence of any property.

Now suppose that all the things 'here' are known to be star-patterned and that a plan is required to bring about a situation in which 'there' contains at least one star-patterned thing. How would such a plan be generated as part of a process of formal reasoning which could be executed by a machine? It turns out that a dozen or so axioms are required to express even these few simple facts, and that the process of proof (that there exists a possible situation in which at least one star-patterned thing is 'there') is far from trivial. Let us list a subset of axioms, chosen deliberately so as to assist construction of a proof. These are expressed in a sloppy information language quite unsuitable for machine use, but convenient for the present purpose of conveying a quick impressionistic sketch of the problem.

1) If a thing is held, then it is impossible for the hand to be in a given place and the thing not to be in that place.

2) If a thing is at a given place then the presence of the hand at this same place means that after doing a pickup a thing-taken will be at that place.

3) If the hand is at a place then any thing-taken will be at that place.

4) After going to a place, the hand will be at that place.

5) After doing a letgo nothing is held.

6) If a thing is held then it is held after the hand goes to a place.

7) If a thing is at a given place then it will be at the place after the hand has gone to the place.

8) If a thing is not at a given place then either the thing is held or it is still not at the place after the hand has gone to the place.

9) If a thing is at a place, then it is at that place after doing a letgo.

10) If a thing is not at a given place then it is not in that place after doing a letgo.

Now comes the statement of the problem:

11) If a thing is 'here' then it is star-patterned.

12) There is at least one thing which is 'here'.

13) If a thing is 'there' and it is star-patterned then this situation is the 'answer' (this states the goal condition and corresponds to requesting the system to find a way of getting at least one star-patterned thing to 'there').

The proof itself comprises altogether 27 statements, of which the final one is: the 'answer' is the situation resulting from the following chain of actions: do a letgo, then go 'here', then a pickup, and then go 'there'.

When one reads through the chain of inferences corresponding to this simple problem a certain mental revulsion is provoked: surely when a child first successfully executes this or that deliberate manoeuvre there can be no question of its having gone through any such mental rigmarole? Three points can be made. First, it is indeed desirable wherever possible to short-cut deductive sequences, replacing parts at least by simulating the task environment and conducting a search in it.

This is perhaps analogous, in human terms, to carrying out trial sequences of actions in the mind's eye as an aid to planning. Those engaged in developing the theorem-proving approach regard theorem-proving not as the be-all and end-all of planning but rather as the default to which the planning process reverts when short cuts are not available.

Secondly, the deductive rigmarole exhibited above looks less formidable if one imagines the different intermediate results to have been arrived at on different previous occasions and individually stored for future use. The importance of storing results of previous calculations and of chaining them together into useful clichés has been emphasised in the context of psychology by G.A. Miller and others, and studied in simple robot simulations by J.E. Doran. This in itself says nothing about how the starting axioms get there. These are facts which have not been deduced from other known facts; instead, they have in some way been abstracted from direct experience of the world. As examples, consider the statements numbered 1–10. Knowledge like 'After doing a letgo nothing is held' must be acquired by inductive generalisation from a wealth of individual experiments. Common observation of infants suggests that they spend much time in such exploratory activity.

Finally, it does seem possible to find ways of uniting the 'planning by simulation' and 'planning by theorem-proving' approaches; if present investigations bear fruit, deductive searches which would otherwise be unbearably cumbersome may thereby be streamlined.

It is fairly obvious that for recognition to advance beyond the mere identification of highly schematised objects using pre-programmed descriptions, powers of generalisation are required. New descriptions, say of 'spoon' must be automatically built up from presentation of a range of examples, and high-level categories built up, for example, of objects found to perform spoon-like functions.

The generalisation of plans is at least as important, and as challenging to the would-be mechaniser. To revert to the Popplestone 'blind hand' example, the successful deduction of a plan for ensuring that at least one star-patterned object ends up 'there' should facilitate the conjecture of a similar plan if the problem were presented again with, say, 'striped' substituted for 'star-patterned', and 'here' and 'there' interchanged throughout. No systematic methods have yet been developed for endowing mechanised planning procedures with this property, but a large and increasing concentration of effort is now being devoted by theoreticians to inductive generalisation, along with a variety of different mathematical avenues. In Edinburgh alone some half dozen approaches are being tried by workers in the Computer Science Department, in the Mathematics Unit and also in our own department. A similar spate of activity is to be observed in some overseas centres. The progress made will determine in a highly critical way the performance of the integrated cognitive systems which are likely to be evolved over the next two or three years. I am here thinking of the differences between tasks such as: 'Find a cup', 'Fetch a pencil', 'Put a ball into a box'; and abstraction from experience of new facts such as (in ascending degree of generalisation): 'Put a box into a ball' implies failure; 'Put anything into a ball' implies failure; 'Put anything into anything solid' implies failure.

The topics discussed here do not of course relate to more than a single specialised corner of machine intelligence research. Even in a single university department such as ours it is only one of several lines of investigation; one should mention in particular mathematical models of memory and of language under study by H.C. Longuet-Higgins, S. Isard and P. Buneman. When a certain stage is reached in the study of recognition and planning, equipping an experimental computing system with a real-world interface becomes a relevant step — and indeed a necessary one if current theoretical formulations are to be put to direct test.

REFERENCES

Barrow, H.G. (1970). The Development of a Real World Interface. Man-Computer Interaction, conference organised by the National Physical Laboratory.
Green, C.C. (1969). Theorem-Proving by Resolution as a Basis for Question-Answering Systems. *Machine Intelligence* 4, 183. Edinburgh: Edinburgh University Press.
Hayes, P.J. (1970). How Long Until the First Intelligent Robot? *European AISB Newsletter* No. 9, 4.
McCarthy, J. and Hayes, P.J. (1969). Some Philosophical Problems from the Standpoint of Artificial Intelligence. *Machine Intelligence*, 4, 463. Edinburgh: Edinburgh University Press.
Michie, D. (1968). Machines that Play and Plan. *Science Journal*, 4, 83.
Nilsson, N.J. (1969). A Mobile Automaton: An Application of Artificial Intelligence Technique. Proc. of International Conference on Artificial Intelligence, 509.

Introductory Note to Chapters 12, 13 and 14

Chapter 12 discusses the interpretation of optical images. In this field the combinatorial explosion spells death to conventional sequential procedures. Hence there exists a pressing need to explore ways to emulate some of the highly parallel architectures employed by the vertebrate eye and brain.

Related aspects of computational complexity are popularised in Chapter 13. The semi-conductor revolution makes it possible even for the home hobbyist to perform small but valid experiments in knowledge engineering. The chapter which follows it also has relevance to this possibility: implementations of 'expert systems' sufficiently compact for desk-top micro-computers are now available. A central feature of 'expert systems' in chemistry, molecular genetics, geology, medicine, plant pathology, robotics, chess and other applications is the man-machine communication of descriptive concepts in the form of patterns. 'Humanisation' of machine-made representations must find a place in future designs.

CHAPTER 12
Teaching a Computer to 'See'

A place for 'seeing machines' is foreseen on the shop floor in tasks like inspecting engineering parts and loading them from one machine to another. It may not after all prove so hard to do what once seemed impossible, namely to design robots which are both smart and cheap. The semi-conductor and microprocessor revolution of the last few years makes even the most expensive of all robotic operations, vision, begin to seem commercially practical.

The cost of a hand-crafted prototype for a 'seeing machine', based on the work of Michael Duff at University College, London (*Computer Weekly*, February 24), and on the programming philosophy developed at Edinburgh by J.L. Armstrong and others, would be about £30,000 in today's prices. This assumes inclusion of a conventional minicomputer to drive the parallel array processor. £30,000 for a complete 'seeing machine' is of course too much. But when we allow for cost-differences as between early models and articles mass produced for the market, and when we also allow for the fact that costs of micro-electronic components continue to fall by about 30% a year, then a programmable 'seeing machine' for £10,000 in today's currency does not seem a remote target.

In robot vision we have to bear in mind that there are two ends to a robot: a front end where the action is, and a back end which ruminates on what the front end and its instructors tell it. Rumination is to a large and unavoidable extent a sequential process. Although expensive, the informational traffic is slow, or should be. Hence the overall cost of, say, matching visual descriptions extracted from the camera with stored descriptions, or planning what pictures to take next, need not be unduly high. The main burden, then, of cost-cutting is thrown on the low-level processes going on in the front end, where every possible trick of special hardware and parallel computing should be pressed into service.

A celebrated paper by Lettvin, Maturana, McCullough and Pitts was called 'What the frog's eye tells the frog's brain'. Their general conclusion was that most of the practical chore of vision, including virtually all the calculation involved in recognising simple edible objects such as bugs, was carried out in parallel array processors in the frog's retina, not by sequential operations in its brain. Recent experience at Edinburgh entirely supports the general conviction that for practical hand-eye robotics the frog's way is the best way.

Visual inspection by computer-controlled TV has until now been a practical impossibility. Billions of separate operations are needed to screen out the noise from a TV image and to identify the simple key features such as lines, corners and bulges from which the final interpretation is constructed. In the past, computers executed their marching orders one step at a time, instead of being able to do, say, 10,000 steps in parallel. The new array processors, like Dr Michael Duff's CLIP developed at University College, London, operate on the parallel principle: the hang-up is 'How do you program such a machine?' A CLIP command language has been designed and implemented. Use of the language for elementary tasks of robot vision has yielded estimated speed-up factors ranging from ten thousand-fold to a million-fold.

Potential developments are not limited to machine inspection of engineering parts. There are moves in the direction of putting large areas of human knowledge — scientific, educational, technological, economic and political — into computer databases from which the decision-taker and expert planner can obtain instant retrieval of needed facts. Anyone who has looked at this problem is struck by the need, not so much for reading machines for getting the knowledge into the machine, since workable solutions for optical input of text to computers are within our grasp, but for diagram-understanding machines. This is the way in which at present we convey much technical knowledge to human assistants.

Finding how to program CLIP-type computers takes a step towards instructing robot assistants by showing them diagrams, whether engineering drawings, chemical structures, chess positions, or whatever, instead of having to spell everything out in Fortran, or in dog-English.

REFERENCES

Duff, M.J.B. and Watson, D.M. (1974). A parallel computer for array processing, in *Information Processing 74*, Amsterdam: North Holland.
Armstrong, J.L. and Jelinek, J. (1977). CHARD user's manual (A Fortran Clip Emulator). Machine Intelligence Research Unit. Research Memorandum MIP-R-115. Edinburgh: Edinburgh University Press.

CHAPTER 13

Artificial Intelligence in the Micro Age

What is Artificial Intelligence? I will give a quirky answer to provoke thought.

Figures 13.1, 13.2 and 13.3 depict three superficially similar games. They are Nim — 'standard' computation given one magic principle; Chess — 'semi-hard'; large catalogue of principles required: Roadblock — for large versions 'hard'; no principles exist.

Nim exemplifies the whole class of problems which can be solved *algorithmically*, i.e. by compact computer programs. At the other end of the scale, Roadblock stands for a class of problems of high *inherent complexity*. Except in trivially small versions, they will never be solved by any computing system.

It might seem that Nim and Roadblock between them cover the whole range but this would be wrong — the real action lies in the no-man's-land between, a territory which I have termed 'semi-hard'. In taking chess to exemplify this category, I should remark that it has not yet been *proved* to differ from Nim. A simple mathematical rule for playing perfect chess *might* still be discovered.

I do not know anyone who believes this. Note, too, that the category exemplified by chess, soluble only by non-compact programs stuffed out with large bodies of knowledge-based rules, is the category which also contains innumerable socially relevant problems of mental skill.

Computing technologies are about to break into this category on a large scale. The question is whether the break-in should be by AI methods modelled on the human style of cognition, or whether it should be by the brute-force technologies of nanosecond processors and trillion bit stores.

Here are three fundamental propositions:

1) No semi-hard problem can be solved feasibly by computer program, unless the program is enriched with a larger or smaller catalogue of logically redundant heuristic information.

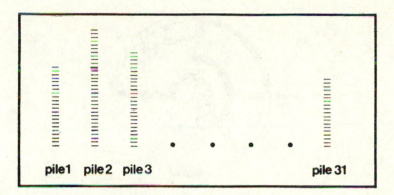

FIGURE 13.1.   A large game of Nim. At the start there are 31 counters in each
pile. Total number of possible positions is similar to chess.

FIGURE 13.2.   Chess: 'A very complicated position.'

2) Solubility can be conferred by a wide variety of such catalogues in
each case but very few will do it so as to preserve human comprehensi-
bility of the program and its operations. Unfortunately the *non*-human
representations and strategies are those which on criteria of machine
efficiency show up as necessarily more cost-effective in action. They are
also, pending radical advance in AI, enormously cheaper to construct.

3) Hence design philosophies of the humanising kind are likely to be
swept aside in the 'advanced automation' rush for economic profit.

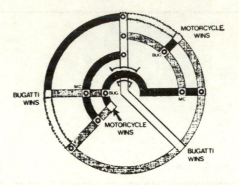

FIGURE 13.3.   A small game of Roadblock. The running time of even the best program increases as the power of the number of intersections. This reflects the fact that no improvement on brute-force search is possible (*see Scientific American* 5 240).

The 1973 mishap at Edinburgh — of which more later — pre-figures in microcosm what may happen in the larger world of technology if we are not very careful.

I fear that the havoc of the first industrial revolution may be repeated on a more uncontrollable scale. To explain my reasons for this fear, let me repeat that there are two ways to solve a semi-hard problem by computer. The 'brute force' way typically gives a 'bigger bang for the buck'. So it will be preferred by an institution with clout.

Slow and Costly

In the Artificial Intelligence way, *human* representations of problem-solving know-how are built explicitly into the program structure, thus preserving the vital link of man-machine comprehension. In the present state of the art 'knowledge engineering' is a slow and costly process. Research directed towards automating it is thus the urgent task. In this country, however, AI tends to be seen as, at best, an esoteric pursuit and, at worst, a shocking expense. There was a time when these snap judgements possessed more than a little truth. They have been once-and-for all unsnapped by the microelectronics tidal wave.

My friend Ed Fredkin, Director of the celebrated MIT Project MAC, asked me recently about I.J. Good's predicted *Ultra-Intelligent Machine*, that ultimate breakthrough when someone will exhibit for the first time a machine — to steal words from the calypso:

'Smarter than the man in every way'.

Fredkin's point was that with computer power as cheap as water it may be a hobby computerist who, in his home, first works the trick.

I do not go at all the way with Fredkin but I think that significant AI work can and will be done by home computer enthusiasts.

Let us consider robotics. Can hobbyists build robots? In the U.S. they not only build them but race them, in the regular *Amazing Micro-mouse Maze Contest*. Yet I regard those robots as boring, bad for all concerned, and a waste of good talent. The nature of the contest, running a simple maze against the clock, encourages intense devotion to sensors, to mechanical ingenuity, to clever circuitry, to cheap software tricks — to everything, in fact, which characterises the runaway technology of the larger world outside, with nothing of the more cognitive attributes which make real mice more interesting than micro-mice.

What technical objectives would be encouraged by the more cognitive type of contest? Humanly and nationally they are not without importance. I believe that it would apply a forcing function to systems which should be to some degree *teachable* and *self-programmable*.

Problem Solving

Since our computing industries are facing a worsening programmer famine, the timeliness and social relevance of such goals cannot be in doubt and I hope for a reversal of an unfortunate prejudice against the discipline in this country.

The first and most drastic manifestation of this prejudice was the withdrawal in 1973 of Science Research Council support for the Edinburgh University *Freddy* AI project. Britain's chances of leading in robotics R and D were put at risk. Robots are not essential to the study of Artificial Intelligence but Artificial Intelligence is as essential to advanced robotics as aerodynamics is to aero-engineering.

How can computers be taught to think? How can they be made 'artificially intelligent'? The first problem to be solved in creating useful AI systems is to teach machines to solve problems by themselves. On its own, however, machine problem-solving is nothing exceptional. Like tic-tac-toe or the '5-puzzle' sliding-block problem, the class of problem may just be too easy to be interesting. Alternatively, the problem may be *very hard indeed*, yet perfectly-executed computer solutions would be not in the least noteworthy.

In chess the defence of king and rook against king and queen, no other pieces being on the board, is so difficult that against Master play of the queen's side there is almost certainly no person alive who can solve it. Moreover, for the queen's side this ending is known to be a theoretical win. So it seems hardly surprising that the task of averting defeat against Master play should be beyond human powers.

Nevertheless, this task can be accomplished by machine-stored expertise, as was demonstrated by Kenneth Thompson at the 1977 meeting of the International Federation of Information Processing in Toronto. He challenged two strong International Masters, Hans Berliner and Lawrence Day, to demonstrate the play of the king and queen's side against a king and rook's defence conducted by his program running on a PDP-11.

Naturally, the two chess masters accepted the challenge, expecting an easy time. To the amazement of onlookers and their own deep mortification, they could make no progress. Time and again new starting positions were set up, but in the ensuing variations the Masters repeatedly lost the thread. When play was finally abandoned the program remained undefeated.

Surely, then, the computer way of solving problems in the KRKQ domain must be very interesting, since it passes with flying colours a gruelling test which chess-masters would flunk. Not at all. The machine's method, however powerful in this task environment, is in itself uninteresting. The machine has memorised a crib.

The total number of legal chess positions in the king-queen-king-rook ending is about three million. So a complete tabulation is possible, in principle, giving for each position the optimal move. Thus if it is a White-to-move position — suppose that the queen's side has the white pieces — a move is entered in the table which lies along the shortest forcing path to checkmate or rook-capture.

If it is a Black-to-move position, the corresponding table entry will contain a move which allows the length of the residual forcing path to be shortened by no more than one move — Black cannot do better than that against best play.

A program for White which looks up its next move in the table is guaranteed to win in at worst the fewest number of moves needed theoretically to force the win, and with any luck, if the defence makes mistakes, in a good deal fewer.

Likewise a program for Black, such as that with which International Masters Berliner and Day had to contend, is guaranteed to spin out Black's demise by the greatest amount possible. If White makes frequent small errors, or infrequent large ones, then the table-stored strategy for Black may survive indefinitely, as the two hapless chess-masters discovered. The longest optimal path consists of 16 moves by White and 15 Black replies.

In remarking that perfectly-executed computer solutions of very hard problems can be nothing noteworthy. I do not imply that what happened at Toronto was uninteresting. On the contrary, it was a gripping, even disturbing, experience for all present. But the interesting phenomenon was not the machine's behaviour, but the *imperfection of Master performance*.

This leads to intriguing questions about its causes, which centre on the drastic resource limitations of the human brain relative to modern computing equipment. The expert practitioner is obliged to package his knowledge into a set of simplifying rules which he can carry in his head, even at the cost of being let down by his rules from time to time.

Unlike the final stored look-up table, the process by which such tables can be computed has something to interest the reader, particularly if he has a home computer with enough store to hold a strategy table for games of non-trivial dimensions.

Let us for convenience illustrate with a trivial one, namely Noughts and Crosses, called tic-tac-toe in the USA. The method of construction is applicable to all games which, like chess, checkers, Go, and five-in-a-row are:

1) two-person;

2) finite — the rules ensure that play must eventually terminate;

3) zero sum — what is good for one player is bad for his opponent to an exactly equal degree;

4) perfect information — both players have sight of the board and the moves;

5) without chance moves — no dice, random draw of cards.

The idea is to start at the end of the game and work backwards. So we must find a procedure for generating all the terminal positions systematically. For a table-based program implementing perfect play for both sides we need construct only two sub-tables, one giving winning play for Nought for all Nought-winnable positions and the other giving Cross strategy for all Cross-winnable positions. If neither Nought nor Cross finds that the current position is missing from both sub-tables, that position is not winnable by either side. The program then has a theoretical draw and must avoid selecting a losing move.

Sub-table Creation

Sub-table creation will be illustrated for the case of building a winning Nought strategy. After eliminating recurrences of the same positions by mirror-imaging or rotation, the number of positions won for Nought can be grouped into 16 Noughts-only configurations, as in Figure 13.4. From each of these, one or more positions can be constructed, according to where the crosses are placed. The corresponding numbers are shown in the right-hand column, making 77 to be stored in all.

The next task is to construct all the possible direct predecessors. They are Nought-to-move positions winnable in one move. They can be obtained by making unit deletions in each position from the three-noughts-in-a-row line. In other words, we ask 'What could have been Nought's last move? Un-make it'. Going back one step further we want to create all the Cross-to-move predecessors which are winnable for Nought. This is more tricky. Let us start with D91.1 as an example.

The first of the steps is straighforward, and yields

```
O   O   X         O   O   X             O   X

O   O   X         O       X         O   O   X

X   X             X   X   O         X   X   O

Nought to move    Nought to move    Nought to move
   D91.1A            D91.1B            D91.1C
```

Next we try to find possible predecessors of D91.1A winnable for Nought — and draw a blank. Cross to move can plug the corner square and win. So, no table entry here. The next case, however, yields a winnable-for-Nought predecessor, and so does the last one, D91.1C.

o	o			o	x	
o		x		o	o	
x	x	o		x	x	o

Nought-winnable   Nought-winnable
predecessor of   predecessor of
D91.1B   D91.1C

The nought-winnable predecessors of *these* are generated by deletion of the 'nought' which participates in both of Nought's potential winning lines, namely the top left corner in D91.1B and the centre of D91.1C. Other nought-winnable predecessors, if the backwards exploration is being done in systematic 'breadth-first' fashion, will be found to have been encountered already and stored during backing-up from other terminal wins.

Proceeding in this way, the strategy-tree is grown backwards until it can be grown no more. Here, then, in essence is our table. The pre-terminal positions are the 'arguments' — to use the language of schoolroom table-look-up — and the moves are the 'values', just as the number 25 entered as an argument in a table of square roots has 5 as the corresponding value.

A few details remain, such as the occurrence of the same position more than once at a given level in the tree. These cases correspond to positions which have more than one equally good winning line.

Figure 13.5 shows another problem, a one-person game. The '5-puzzle' shown in the Figure 13.5 is a poor relation of the formidable '15-puzzle'. The latter, incidentally, is discussed in the last published writing of A.M. Turing, the mathematical logician and pioneer of computing.

As far as its mathematical properties are concerned, the 5-puzzle has been polished-off in a page by P.D.A. Schofield, who points out an amusing correspondence between the classes of move-sequence in the 5-puzzle and the axes of symmetry of the dodecahedron — the twelve-faced

	1	2	3	4	5	6	7	8	9	10	Totals
D51											6
E51											7
M51											6
D71											10
D72											5
E71											9
E72											5
E73											9
M71											9
M72											5
D91											1
D92											1
D93											1
E91											1
E92											1
M91											1
											77

FIGURE 13.4.   Tabulation of all legal Noughts-and-Crosses positions which have been won by Nought, the opening player. D, E and M correspond to 'diagonal', 'edge', and 'middle' winning patterns.

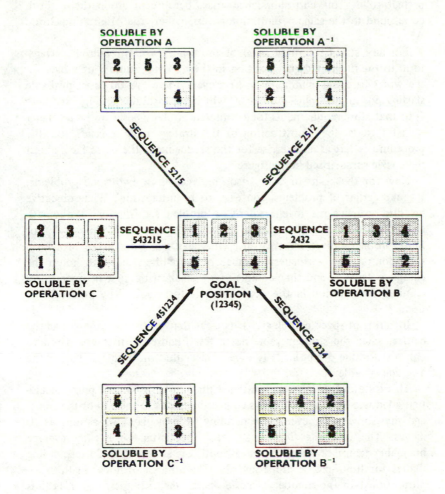

FIGURE 13.5.   The four positions soluble in four moves and the two positions soluble in six moves have been arranged and labelled to explain the notation used in the strategy table for the six basic operations A, A⁻¹, B, B⁻¹, C, C⁻¹. The first four operations correspond to move sequences of length 4, the last two to move-sequences of length 6.

perfect solid. This correspondence arises because in group theory both correspond to the same permutation group, namely the Alternating Group A5.

For any starting configuration of the 5-puzzle it is required to transform to the illustrated goal position in the minimal number of moves. In the worst case this minimum is 20 moves. There are 60 distinguishable starting positions — plus another 60 which are insoluble and ignored here — so that store requirements for a complete strategy-table are even smaller than for tic-tac-toe. Construction of the strategy table is easier, too; the procedure is left as an exercise for the reader using the basic facts about the puzzle summarised in the figure.

Now for the point of comparison between these two small problems. I spoke earlier of problems 'too easy to be interesting'. What objective measures can we use to characterise a problem's difficulty? Mathematicians have devised such measures, and they speak of 'complexity'. For finite problems there are two different measures, namely the problem's space-complexity — smallest number of store-bits needed to house the complete table — and time-complexity — smallest number of operations' required for solution in the case that it is to be solved by pure calculation.

In terms of space complexity it is clear that tic-tac-toe emerges as the harder, since the strategy table has a few hundred entries compared to only 60 for the 5-puzzle. It is equally clear that on pure calculation the 5-puzzle is harder.

This follows from the circumstance that the only known pure calculation adequate in all cases follows all possible paths to the end of the game; and that some paths, even corresponding to 'best play', are as long as 20 moves. The complete 'lookahead tree' for tic-tac-toe has an average branching factor of about 3 and a depth of, at most, 8. Corresponding figures for the 5-puzzle are 2 and 20. The number of nodes in the tree, proportional to the number of basic operations to be performed, is thus 3^8 and 2^{20} for the two cases, i.e., 7000 and 1,000,000 respectively. So on time complexity the 5-puzzle is the harder.

The magic trick on which human culture and intellectual history has been built, the trick of cognition, lies in ingenious compromise. Each of the two approaches is of impractical cost by itself. But the right blend can shrink costs miraculously.

TABLE 13.1

	Condition-Pattern	(Constraint)	Goal-Pattern
Rule 1	total-distance = 0	(NIL)	HALT
Rule 2	preferred edge-pair 0 apart & edge-pair-distance = 0	(A, A^{-1}, B, B^{-1})	total-distance = 0
Rule 3	preferred edge-pair 0 apart & edge-pair-distance > 0	(C, C^{-1})	total-distance is reduced
Rule 4	preferred edge-pair 1 apart & intervening piece at place 2	(A, B^{-1})	edge-pair-distance is reduced
Rule 5	preferred edge-pair 1 apart & intervening piece not at place 2	(A^{-1}, B, C, C^{-1})	distance-of-intervening-piece-from-place 2 is reduced
Rule 6	both edge-pairs 3 apart	(A^{-1}, B)	not (both edge-pairs 3 apart)

Note: A Constraint is applied as follows. Each operation in the bracketed list is applied in turn to the current position and the resulting position is checked against the Goal-Pattern. When a match occurs, the successful operation is selected and the Table is re-entered as Rule 1.

Pattern-directed 'advice program' for the 5-puzzle generates near-optimal solutions at miniscule cost of store and predecessor. The rules are taken in order, and the first whose condition-pattern matches the current state of the problem is executed. Each rule is interpreted in the style; *given* that the condition-pattern has been matched *then* (using the constraint to reduce needless move-trials) find a move-sequence which creates a match with the goal-pattern and *then* apply the sequence *and then* re-enter the table.

Bundles of rules processed according to this regime are often referred to in AI work as 'production systems' owing to their resemblance to a formal scheme developed by the celebrated logician E. Post. A neat and eminently revisable framework is thus provided for packaging useful heuristic information about the problem, and keeping the resultant 'knowledge-base' quarantined from the 'knowledge-interpreter', which does the heavy-duty computation involved in search, pattern-matching and the like. Such knowledge-bases can be treated as data and edited by program – 'machine learning' – to improve performance, or they can be modified or augmented interactively by a human expert in the given problem-domain, thus exhibiting the desirable property of 'teachability'.

The important thing is that the level at which the problem is conceptualised in such sets of pattern-based rules should correspond closely to the human's mental picture and thus lend itself to reciprocal transfer of knowledge between user and problem-solving system.

What does this blend look like? Let me phrase the same question in the form of a table as follows:

Approach	Pros and cons
1) A try-everything program	Little store needed, but runs for ever
2) A compromise between a memory-intensive and a calculation-intensive representation	Store and computation requirements both moderate
3) A situation-action dictionary	Little computation but store would fill the world

The second approach listed is a dictionary of a kind but instead of zillions of situations entered separately they are grouped into a smaller number of situation-*types*. Instead of each entry giving an action as a result, an action *scheme* is entered. A scheme is some more general picture — a set of goal-patterns and constraints — from which actions can be recovered by calculation.

So the new kind of dictionary is a dictionary of *patterns* in place of individual instances. Each entry, in effect, says to the processor: 'If the present situation matches *this* pattern, then see if you can work out a way of creating a new situation matching one of the following *target patterns*, using only those actions which match these *constraint patterns*?'

What this might look like for the 5-puzzle is shown in the pattern-directed strategy-table from which we recover general pieces of *advice* rather than individual moves. Such a 'knowledge-base' of pattern-directed rules might be the only way of getting a strategy into a machine if the latter were a hand-held programmable lacking enough memory for the complete 60-entry exhaustive tabulation. Needed definitions are given in Table 13.2.

The table's strategy follows a well-worn approach often called 'problem-reduction', whereby the goal is decomposed into constituent features to be tackled separately. For the 5-puzzle our strategy sets the sub-goal 'solve an edge' and then proposes solving the residual problem.

It is common, as in the present case, for such a sweeping simplification to sacrifice the guarantee of optimality in the solutions generated but if the solutions are near-optimal the sacrifice may be judged worthwhile. Table 13.3 gives the results of running the 'advice program' sketched in Table 13.1 on an LSI-11 micro, coded in BCPL.

This last consideration comes to life in the present context as soon as we realise that the sub-goaling ruse adopted generalises to sliding-block puzzles in general. Moving to the 8-puzzle — 3 × 3 board, central square by convention empty in the goal configuration — problem-reducing takes the forms, solve one of the four edges, and then solve for the residual 5-puzzle.

Sufficient Challenge

Clearly the technique can be pushed higher and higher, to the 11-puzzle, the 15-puzzle, the 19-puzzle, and so on. For a home micro owner, however, I would suggest that a sufficient challenge initially would be to extend to the 8-puzzle the methods illustrated, using Schofield's paper already cited as general background.

Best of all is the kind of program which has so general a structure that different knowledge bases can be slotted in and out according to which puzzle is to be tackled.

If we raise the scale of problems to the level of chess and the scale of solving device to the level of the trained human brain, we can see the answer to an otherwise puzzling riddle. Since pure search takes us nowhere in such huge problem domains, and since a complete strategy table is also not a thinkable proposition, how does the chess-master find good moves?

Investigations by Alfred Binet at the turn of the century, by Adrian de Groot in the years after the second world war and by Herbert Simon and colleagues more recently, all point to the same trick. It is based on amassing pattern-based mental catalogues of the same essential kind as the table of advice illustrates for the toy example of the 5-puzzle.

Other work suggests similar conclusions for skilled intellectual know-how in general, whether in medical diagnosis, plant pathology, chemical compound identification and synthesis planning, or decision-taking in geological prospecting. Computing systems capable of this kind of practical thinking in specialist areas of applied knowledge are called 'expert systems'.

TABLE 13.2
Definitions required to implement the strategy of Table 13.1, results obtained
with a version in BCPL for the LSI-11 micro.

---

total-distance: the sum of the five *piece-distances*, where

piece-distance = the shortest number of moves required to get the given piece home if all other pieces are removed from the board — equal to the sum of the absolute values of the x- and y-coordinate differences between present location and home location.

edge-pair: the piece-pair (5, 1) or the piece-pair (3, 4). In the goal configuration these two pairs occupy the left-hand edge and right-hand edge respectively.

edge-pair-distance: the sum of the *piece-distances* of the two members of the *edge-pair*

preferred edge-pair: the *edge-pair* with fewer intervening pieces; in case of a tie, then the *edge-pair* with the lesser *edge-pair-distance*, if still tied, then choose arbitrarily.

edge-pair n apart: starting with piece 5 (or 3 as the case may be) proceed clockwise round the board counting the intervening pieces until piece 1 (or 4 as the case may be) is reached.

place 2: the location on the board occupied in the goal position by piece 2.

Once the essential principle has been grasped, it is within the resources even of the micro hobbyist to build interesting small systems of this type. The key principle is that programs must be written in a new way, namely in the form of modular and incremental bundles of pattern-based rules.

Rules are invoked by processes of matching with the current state of the problem rather than by explicit sub-routine call. The ability of pattern-directed programming to steer between the clashing rocks — between the Scylla of processor-exhaustion and the Charybdis of store-exhaustion — is as fundamental to the success of today's expert systems as the phenomenon of aerodynamic lift was to the pioneers of heavier-than-air flight.

It may be thought that something has been said to illuminate the nature of human cognition. Certainly these machine models, the so-called 'expert systems' throw light on one particular aspect of cognition — the use of the brain for routine execution of acquired skills.

Although this is the cognitive mode in which most of us spend the greater part of our waking lives, it occupies a fairly lowly rung on the ladder of the intellect. The next rung up is the ability autonomously to *acquire* pattern-directed skills. In learning from precept, from example and

TABLE 13.3
Frequency distributions of lengths of solution-paths for the 5-puzzle.

Position	Program	Optimal	Position	Program	Optimal
12345	0	0	34125	16	14
12453	10	10	34251	10	10
12534	10	10	34512	12	12
13254	14	14	35142	16	14
13425	4	4	35214	20	16
13542	14	14	35421	8	8
14235	4	4	41253	16	14
14352	14	14	41325	14	14
14523	16	14	41532	14	12
15243	20	14	42135	14	12
15324	14	14	42351	10	10
15432	18	18	42513	16	16
21354	14	14	43152	20	16
21435	20	18	43215	18	16
21543	18	16	43521	14	12
23145	16	16	45123	12	12
23451	6	6	45231	10	8
23514	16	14	45312	16	14
24153	20	16	51234	6	6
24315	20	14	51342	4	4
24531	10	8	51423	10	8
25134	10	10	52134	14	12
25341	4	4	52314	10	10
25413	14	12	52431	24	20
31245	16	16	53124	10	10
31452	10	10	53241	20	18
31524	20	16	53412	10	8
32154	24	20	54132	8	8
32415	14	12	54213	14	12
32541	14	12	54321	18	18

from practice, only modest progress has been so far made by the mechanisers.

Above this lie regions of creative insight and the higher flights of abstract reasoning. Machine systems for these higher levels still lie in the
future. It may be apposite to close with a problem due to John McCarthy
which is intractable to pure search, to pure table look-up, and to all

mixtures and blends of the two. Yet it falls apart when the right insight is brought to bear.

The problem is posed in two stages. The first is trivial. Can a checker-board be covered neatly with 32 dominoes, each domino being of a size exactly to cover two adjacent squares? Obviously, yes. Now cut off the top left and bottom right squares of the board. Can the multilated board of 62 remaining squares be tiled with 31 dominoes?

If you think that your program might be able to slug it out by trial-and-error exhaustion of possible domino-tiling patterns, then I merely multiply the board's dimensions by 10, so that it has 6,400 squares, and declare the essential problem unaltered.

Finally someone points out that each domino covers exactly one white and one black square. Initially there are equal numbers of the two colours; but two opposite corner squares of an even-sided board must be of the *same* colour, say white, so that their removal creates a surplus of two black squares remaining at the end. Hence the even-sided mutilated checker-board cannot be tiled.

REFERENCES

Schofield, P.D.A. (1967). Complete solution of the "Eight-Puzzle", *Machine Intelligence 1* (eds. N.L. Collins and D. Michie) Edinburgh: Edinburgh University Press.

Thompson, K. (1979). Walter Browne v. Belle, May issue of *British Chess Magazine*, gives the results with commentary of two challenge games between a strong Grandmaster and the KRKQ database.

Turing, A.M. (1954). Solvable and unsolvable problems. *Science News*, London, Penguin, pp. 7-23.

CHAPTER 14

Expert Systems

A new programming technology has been growing up around the problem of how to transfer human expertise in given domains into effective machine form, so as to enable computing systems to perform convincingly as advisory consultants. Expert systems development, confined during the past decade to academic laboratories, is becoming cost effective. Reasons are partly advance of semiconductor technology and partly development of well understood methods for 'knowledge-based' programming.

A few examples may be in order to illustrate the kinds of consultant skills under discussion (Table 14.1). MOLGEN (Martin *et al.*, 1977) interactively aids molecular geneticists in the planning of DNA-manipulation experiments. VM ('Ventilator Management'; Fagan, 1978) gives real time advice for the management of patients undergoing mechanical ventilation in the intensive care of the Pacific Medical Center. PUFF (Kunz, 1978) interprets results of pulmonary function tests in use in the same centre. SACON (Bennet *et al.*, 1978) guides engineers in the use of a large program which integrates structural analysis procedures. PROSPECTOR (Duda *et al.*, 1979) advises when and where to drill for ore. DENDRAL (Buchanan, 1979) takes the pattern generated by subjecting an unknown organic chemical to a mass spectrometer, and infers the molecular structure. SECS (Wipke, 1974) uses a 'knowledge base' of chemical transforms to propose schemes for synthesising stated compounds. End-game expert systems deploy and discuss chessmaster knowledge and generate improved teaching texts (Bramer, 1980; Bratko, 1978; Bratko and Michie, 1980). MYCIN (Shortcliffe, 1976) and INTERNIST (Pople *et al.*, 1977) outperform clinical consultants within the limited domains of expertise of these programs. Further information on many of these, and on other recent systems, is available (Michie, 1979). Practical insights obtained so far can be summarised as follows.

195

TABLE 14.1
Examples of expert systems.

Medicine	Identification of bacteria in blood and urine samples, and prescription of antibiotic regime	MYCIN (Shortliffe)
	Diagnosis in internal medicine	INTERNIST (Myers and Pople)
	Intensive care ('iron-lung')	VM (Fagan and others)
	Interpretation of lung tests	PUFF (Kunz)
Chemistry	Identification of organic compounds	DENDRAL (Feigenbaum, Lederberg, Djerassi, Buchanan, Carhart and others)
	Designing organic syntheses	SECS (Wipke)
	Molecular genetics	MOLGEN (Lederberg, Martin, Friedland, King, Stefik)
Other	Consultancy for structural engineers	SACON (Bennett)
	Consultancy for mineral prospecting	PROSPECTOR (Hart, Duda, Einaudi)
	Chess end-game play and advice	AL1, ALI-5, and other advice programs (Michie, Bratko, Niblett, Bramer)

1) The market for consultancy demands specialists, not generalists; this applies to automated consultancy too.

2) Real time operation is in some applications not just desirable but essential (see the reference to VM earlier).

3) A consultant's skill consists to an important degree in asking the client the right follow-up questions, as the outlines of the case takes shape.

4) Unless the program can do this, and can also explain its steps on demand, client confidence suffers.

5) An expert system acts as a systematising repository over time of the knowledge accumulated by many specialists of diverse experience. Hence it can and does ultimately attain a level of consultant expertise exceeding that of any single one of its 'tutors'.

6) Program text in the ordinary sense is an unsuitable and unpopular medium for the description and communication by human experts of their expertise. 'Advice languages' are needed.

Particular attention has been directed to consideration 4 above by Feigenbaum (1979): '. . . The Intelligent Agent viewpoint seems to us to demand that the agent be able to explain its activity; else the question arises of who is in control of the agent's activity. The issue is not academic or philosophical. It is an engineering issue that has arisen in medical and military applications of intelligent agents, and will govern future acceptance of AI work in application areas.'

Nature of Knowledge-Based Expert Systems

Expert systems are not, and owing to the complexity of their tasks cannot be, *either* procedure-driven in the ordinary sense *or* data-driven, although they can all be fairly described as data base-driven. The great bulk of the data base, however, is typically made up of *rules* which are invoked by pattern-match with features of the task-environment and which can be added to, modified or deleted by the user. A data base of this special type is ordinarily called a 'knowledge base', and its existence determines that there are three different user-modes for an expert system in contrast to the single mode (getting answers to problems) characteristic of the more familiar type of computing:

a) getting answers to problems – user as client;

b) improving or increasing the system's knowledge – user as tutor;

c) harvesting the knowledge base for human use – user as pupil.

Users of an expert system in mode (b) are known as 'domain specialists'. It is not possible to build an expert system without one.

In the early days it was thought that computer based consultation would be confined to the conventional performance goal of earlier computing technologies, namely to deliver good answers to the client's input questions. This turned out not to be the whole picture.

First: the client demands explanations as well as answers.

Second: the system is typically required to have facilities for improving and refining its knowledge from tutorial interaction with the client's own domain-specialists.

TABLE 14.2
Excerpt from a consultation with the MYCIN program.

. . .22. Was any chemotherapy recently begun specifically to combat
ORGANISM-1?
. . . NO
23. Is Jeremiah Sample (PATIENT-1) a burn patient?
. . . RULE
RULE047
IF: 1. THE SITE OF THE CULTURE IS BLOOD, AND
       2. THE IDENTITY OF THE ORGANISM IS NOT KNOWN
       WITH CERTAINTY, AND
       3. THE STAIN OF THE ORGANISM IS GRAMNEG, AND
       4. THE MORPHOLOGY OF THE ORGANISM IS ROD, AND
       5. THE PATIENT HAS BEEN SERIOUSLY BURNED
THEN: THERE IS WEAKLY SUGGESTIVE EVIDENCE (.4) THAT
       THE IDENTITY OF THE ORGANISM IS PSEUDOMONAS
. . . NO
24. Is the patient's illness with ORGANISM-1 a hospital-acquired
infection?
. . . YES
25. Do you consider Jeremiah Sample (PATIENT-1) to be a
compromised host? (alcoholic, immunosuppressed, etc.):
. . . NO
26. Has Jeremiah Sample (PATIENT-1) had any gu surgery or
manipulation of the urinary tract? (e.g. indwelling catheter,
cytoscopy, etc.)
. . . WHY
I AM TRYING TO FIND OUT ABOUT THE IDENTITY OF THE
ORGANISM.
I ALREADY KNOW THAT:
THE PORTAL OF ENTRY OF THE ORGANISM IS UPPER-GU
THE MORPHOLOGY OF THE ORGANISM IS ROD
THE STAIN OF THE ORGANISM IS GRAMNEG
THE SITE OF THE CULTURE IS BLOOD
THEREFORE, IF:
1. THE PATIENT HAS NOT HAD A GENITO-URINARY MANIPULATIVE
PROCEDURE, AND
2. URINARY-TRACT-INFECTION IS NOT A PROBLEM FOR WHICH
THE PATIENT HAS BEEN TREATED
THEN:
THERE IS SUGGESTIVE EVIDENCE (.6) THAT THE IDENTIFY OF THE
ORGANISM IS E. COLI
(RULE156)

Third: an additional feature has come to the fore, whereby the system can be made to generate improved codifications of domain-specific knowledge for human use. We call this new craft 'knowledge refining'.

A typical explanation facility allows the user the challenge and examine the reasoning process underlying the system's answers. The dialogue shown in Table 14.2 is excerpted from a session with MYCIN, a program knowledgeable about diagnosis and antibiotic therapy in certain classes of infectious diseases. This record contains two examples of the system's ability to handle questions of the form 'how did you work that out?' In the first, the user types 'RULE' and receives an English-language version of the last rule to be executed. In the second example the user-command 'WHY' triggers a backward trace of the inference process which fired the system's last question. The number of the last rule is also given in case the user wishes to retrieve and examine it. Further backward tracing could be activated by repeated use of 'WHY' and 'RULE' commands before proceeding with the main dialogue.

The supporting software framework is of a type normally called a 'production system' (Davis and King, 1977) — a modular collection of rules, together with a control structure. Each rule has a condition part consisting of a conjunction of patterns C1, C2, etc. paired with an action part (A1, A2, etc.) according to the general scheme shown in Table 14.3. The list of rules is searched for the subset whose condition parts are satisfied ('matched') by the current state of the data base. The retrieved candidate set is processed to detect any conflicts and to resolve them by elimination of rules from the candidate set. The first rule of the reduced set is executed. An action part can be an action, e.g. 'print disease-name' or a logical, numerical or other value, or it can be an action sequence or an action scheme, goal-list or other advice structure used to guide an action generating module. Typically execution of the action part of a rule modifies the state of the data base.

Rule-based Inference

The deductive inferences performed by MYCIN in the process of answering the user's questions follow a control scheme known as 'backward chaining'. Consider a simple set of rules in which letters from the alphabet have been substituted for 'facts'.

TABLE 14.3
Production system 'recognise-act' cycle. At the given instant the data base contains
the system's model, in the form of an implicit conjunction of conditions, of the state
of the task environment. '(C1 & C2) → A1' means '*if* the condition (C1 & C2)
matches the data base, *then* execute A1'. Conflict resolution is the task of a tie-
breaking algorithm, not specified here.

DATABASE: C5 C1 C3 PRODUCTION RULES	RECOGNISE CONFLICT SET	SELECTED RULE
		Conflict
(C1 & C2) → A1 Match		Resolution
C3 → A2	C3 → A2	C3 → A2
(C1 & C3) → A3	(C1 & C3) → A3	
C4 → A4	C5 → A5	
C5 → A5		
	ACT	
	Execution	
C3 → A2		A2 executed

1. A & B → F
2. C & D → G
3.   E → H
4. B & G → J
5. F & H → X
6. G & E → K
7. J & K → X

The arrow '→' implies 'THEN', thus the first rule reads

'IF A is true AND B is true THEN F is true'.

Suppose that in a particular case we discover by observation that 'facts'
B, C, D and E are 'true', and we wish to discover if X is therefore true.

The program will consider those rules which could be used to infer the
truth of X, i.e. those rules (5, 7) which have an X on the righthand side of
the arrow. Each such rule is tested to see if each of the facts on the left-
hand side are known to be true, any unknown fact being treated in the
same way as the original fact X — i.e. we proceed by recursion. Thus:

X may be deduced from Rule 5 or Rule 7
Starting with Rule 5, are F & H true?
  F can be shown to be true if A & B are both true (Rule 1)
    A is not known to be true, so this attempt fails
Continuing with Rule 7, are J & K true?
  J can be shown to be true if B & G are both true (Rule 4)
  B is known to be true *a priori*
  G can be shown to be true if C & D are both true (Rule 2)
    C is known to be true *a priori*
    D is known to be true *a priori*
   therefore G is true
  therefore J is true
  K can be shown to be true if G & E are both true (Rule 6)
    G is already known to be true
    E is known to be true *a priori*
   therefore K is true
  therefore X is true.

The above simple deductive technique is the basis of MYCIN's reasoning. The technique is powerful and efficient while at the same time general and easy to comprehend.

'Learning' expert systems

The rule-based structure of expert systems facilitates acquisition by the system of new rules and modification of existing rules, not only by tutorial interaction with a domain-specialist but also by autonomous 'learning'. An early example was a self-taught pole-balancer developed by Michie and Chambers (1968) on the basis of 225 condition-action rules. De Dombal's diagnosis program acquires its medical expertise by statistical induction over patient records with confirmed diagnoses (de Dombal *et al.*, 1972). Michalski's AQ11 program (Larson and Michalski, 1977; Michalski, 1978) acquires diagnostic expertise by logical induction as also is the basis, following different formalisms, of the successes scored for machine learning in chemistry by Meta-DENDRAL (Buchanan *et al.*, 1976) and in robot vision by the Edinburgh FREDDY system (Ambler *et al.*, 1975). Finally Quinlan's (1979) latest version, ID3, of Hunt's CLS algorithm recently synthesised inductively in a few seconds of machine time solutions to

TABLE 14.4

'Learning' expert systems. The critical factor affecting the speed at which the new technology can take off is the degree to which expert systems can be made capable of adding to, or refining, their own rules in the light of observations — either presented by the user or sampled directly. PROSPECTOR's use of the Bayesian component could justify addition of this program to the above list.

Balancing a pole	BOXES program
	(Michie and Chambers)
Diagnosis of acute abdominal pain	Bayesian decision program
	(de Dombal, Gunn)
Diagnosis of soybean diseases in Illinois	AQ11 inductive inference program
	(Chilausky, Jacobsen, Michalski)
Spectroscopic identification of	Meta-DENDRAL
keto-androstranes	(Buchanan and others)
Visual recognition of parts	FREDDY experimental robot
for 'hand-eye' assembly	(Ambler, Barrow, Burstall, Brown,
	Crawford, Michie, Popplestone, Salter)
Recognition in king-rook-king-knight	PASCAL implementation (ID3) of
endings of 'knight's side lost in 3-ply'	Hunt's 'concept-learning system'
positions	(Quinlan)

classification problems which had proved intractable as tasks of hand-synthesis and coding. A connection thus appears between machine learning and automatic programming. This connection gains interest from the fact that recent runs of ID3 (Quinlan, personal communication) have synthesised programs (in the form of decision trees) which perform classification tasks more than five times faster than the best hand-coded program. Various 'learning' expert systems are listed in Table 14.4.

The system for soybean diagnosis (Chilausky *et al.*) shown in the figure starts with primitive descriptors from the expert pathologist and from these, and from a training set of values for diseased plants with confirmed diagnoses, synthesises a set of diagnostic rules. The unexpected discovery was made that a machine-synthesised set of rules out-performed those developed by the plant pathologist, Dr Jacobsen, who acted as domain-specialist and was the source of the original set of primitive descriptors. Jacobsen then attempted to improve his rules, and partially succeeded as shown in the bottom line of Table 14.5. Feeling that further improvement would be hard, he discontinued the attempt and adopted instead the machine-synthesised set as the basis of his subsequent professional work.

One way of summarising the relation between inductive 'concept

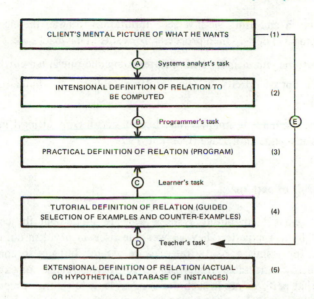

FIGURE 14.1. A way of looking at the synthesis of a concept (description) in the form of a program. The fundamental ideas underlying this diagram are (i) that both the teacher and the programmer have the task of conveying concepts to target devices which are then called upon to apply the acquired concepts to new data and (ii) that a symbolic definition is not the only kind of definition which could be used as a formal specification from which to build a program. A sufficient set of tutorial instances could do the same job. In the case of the teacher's task, the 'target devices' are of course his or her human pupils — providentially equipped with rather good inductive capabilities. Current research aims to equip the programmer's target devices in something like the same way.

learning' and automatic program synthesis is diagrammed in Figure 14.1. An unexpected sidelight on future uses for inductive learning, additional to the obvious ones, is cast by the following consideration. As memory continues to get cheaper faster than processing power, the possibility of encoding industrially useful information in the form of giant look-up tables will begin to be realised in commercial practice. In many cases the time-complexity of the function to be represented in the table makes it unfeasible to initialise such tables in the obvious way. When, however, the function has a low-complexity inverse (as for example the prime factor function or the function mapping from mass spectra to molecular structures) it is possible to initialise such tables 'backwards', i.e. by enumerating f's y-domain and using the inverse computation to enter the elements of the x-domain. Look-up then proceeds 'forwards'. Drawbacks are:

1) cluttering up memory with uninteresting and unwanted entries;

2) conceptual opacity of the resulting table to the human domain-specialist.

Inductive inference techniques have been used for combatting 1 in a study of chess end-game knowledge (Quinlan, 1979).

Critical role of patterns

Knowledge-based computing systems seek to implement the consulting skills of human experts. They answer questions in problem domains too complex for 'standard' hardware/software designs, but not so complex as to be totally intractable. Study of the cognitive strategies of experts has shown that performance in such domains, at least for human practitioners, is not based on elaborate calculations but on the mental storage and use of large incremental catalogues of pattern-based rules. Thus chess mastership is gained through the acquisition and organisation in memory of diagnostic patterns, not through increases in calculating power. In Figure 14.2 the upper two patterns illustrate thematic categories of the sort found in the early pages of a chess primer, 'fork' and 'back-rank mate' respectively (Zuideman, 1974). The lower two exhibit a single pattern differing by a minor perturbation which happens to be critical. In the lefthand case a familiar type of sacrifical attack on the King can be launched with impunity. In the righthand position it can be spiked at the last minute by the move B-Q6 by Black, guarding White's intended Q × RP. The role of remembered patterns is thus to *propose* a tactical idea. Detailed check-out by concrete analysis is still required.

In Table 14.6 some representative pattern-based skills are listed, for four of which 'expert systems' have been implemented. The approximate number of patterns required for successful machine implementation is thus in these four cases known. The last line contains estimates for a highly sophisticated domain of human expertise where no comparable machine solutions yet exist. Table 14.7 shows the approximate numbers of patterns required for a few of the fragments of the total chess domain for which machine mastery has been achieved. These small subdomains could of course be (and have been) solved by brute force enumeration. But this

TABLE 14.5
Experiment by R. Chilausky, B. Jacobsen and R.S. Michalski (1976).

AQ11 in PL1	120K bytes of program space
Soybean data:	19 diseases
	35 descriptors (domain sizes 2-7)
	307 cases (descriptor-sets with confirmed diagnoses)
Test set:	376 new cases
	> 99% accurate diagnosis with machine rules
machine runs using	83% accuracy with *Jacobsen's rules*
rules of different origins	
	93% accuracy with interactively *improved rule*

approach yields representations which cannot support the intelligent query and explanation facilities demanded by the user of knowledge-based systems.

Patterns in computer vision

In chess and other deterministic combinatorial domains (such as industrial routeing and scheduling in various OR contexts) the power of patterns is revealed in the extraction of sense from otherwise intractable explosions of combinatorial complexity. Table 14.7 gives a hint of how well chosen pattern-sets can serve this function, and shows that a relatively slow growth of the pattern catalogue can maintain control over a wildly growing problem space. In some other perceptual domains, notably vision, combinatorial complexity is compounded by the presence of sensory noise, thus putting an even higher premium on the stored pattern-base. Even without low level noise, perturbations can be severe, R.L. Gregory (1970) has a photograph such that if viewed upside-down or on its side when first encountered, there is little chance that the human eye and brain would 'see' anything but a formless chaos of blotches. Yet when re-oriented it is clearly seen as a photograph of a Dalmatian dog drinking from a puddle in a stone-strewn landscape. The feat whereby sense is extracted from noise rests on the fact first emphasised by Helmholtz in the 19th century that visual perception is an act of *reconstruction* of the

FIGURE 14.2.   Some 'patterns' in chess (from Zuidema, 1974).

percept from a large repertoire of stored internal models. The rate of input of visual information to the higher centres of the brain is not great enough to do more than give hints and prompts for the reconstructive process. We catch the mechanism in the act whenever we 'see' in randomly blotched surfaces pictures which are not 'really' there — 'similitudes of all sorts of landscapes and figures in all sorts of actions' as Leonardo da Vinci said of these apparitions.

Removal of noise is executed in two main phases — a pre-processing phase in which knowledge of an essentially statistical nature is applied for smoothing and cleaning up the raw picture, and a second phase,

TABLE 14.6
Some pattern-based skills (condensed from Michie, 1979a).

Skill	Nature of implementation	No of pattern-based rules in implemented system
Seeing a scene	Incremental catalogue of visual patterns. Simple scenes of shadowed polyhedra. (Waltz, 1972)	10
Balancing a pole	Incremental catalogue of pattern-based rules. (Michie and Chambers, 1968)	225
Identifying organic compounds from mass spectra	Incremental catalogue of pattern-based rules. 'Dendral' program of Lederberg, Feigenbaum and Buchanan	c400
Identifying bacteria from lab tests of blood and urine	Incremental catalogue of pattern-based rules. 'Mycin' program of Shortliffe (1976)	c400
Calculating-prodigy arithmetic	Alexander Aitkin studied by Hunter (1962) used pattern-based rules	?
Grandmaster chess	Chessmasters, studied by Binet (1894), de Groot (1965), Chase and Simon (1973), Nievergelt (1977) use pattern-based rules	30,000–50,000

which the dog photo was selected to illustrate, in which semantic knowledge comes into play. For handling knowledge of the second kind a commonly used representational form is the *semantic net*, of which Figure 14.3 shows an early example. In the context of machine expertise in vision, the notion of 'internal models' adapted by Gregory from Helmholtz thus receive specific and concrete realisation.

There is a practical bearing of computer vision on expert systems work, owing to the need from time to time to resort to diagrammatic explanations and other 'picture talk' in the course of man-machine consultations. If a medical program is advising on a case of acute abdominal pain it would be advantageous to be able to input the standard diagram of the abdomen from the patient's notes, filled in graphically to indicate regions

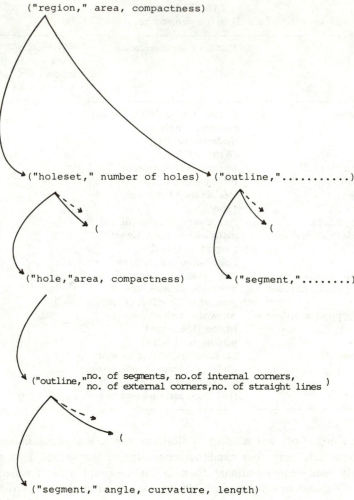

FIGURE 14.3.   The diagram depicts the descriptive structures used in the Edinburgh robot project (Ambler *et al.*, 1975) to encode diagnostic information about solid objects viewed by the computer-controlled robot through a TV camera. The slots in these structures could only be filled after a variety of pre-processing routines had acted to eliminate noise in the picture, to identify optically homogeneous regions, to find, trace and segment boundaries, and to perform various measurements on the primitive features thus isolated.

TABLE 14.7
Pattern requirement of three small subdomains of chess grows slowly relative to
increase of domain complexity.

Ending	Size of the problem space	Number of patterns required for an expert system
King and Rook against King (Bratko, 1978)	40,000	10
King and Pawn against King (Bramer, 1980)	100,000	20
King and Knight against King and Rook (Bratko and Michie, 1980)	2,000,000	30

of tenderness, rigidity, etc. rather than to have to construct symbolic circumlocutions. Beyond a certain level of complexity e.g. in computerised fault-diagnosis in the production machinery of an oil platform, the task of circumlocution can become intractable.

Past work towards supplying such needs has until recently been rewarded by lack of highly parallel hardware designs. Computer visions can certainly use these. Working with a FORTRAN emulator of Duff's CLIP-3 parallel array processor, Armstrong and Jelinek (1977) utilised a command language for vision in which they were able to specify solutions to the normal range of low level vision tasks — removing noise, finding and measuring blobs, following lines, detecting vertices and so on. Although emulator-overheads slowed their algorithms down by factors ranging from a thousandfold to ten-thousandfold, they were still beating standard sequential algorithms in real time. The reason, it turned out, was that without a language in which to express parallelism it is not easy to acquire the mental set for seeing simple fast ways of doing these things. In the next generation of knowledge-based systems, incorporation of versatile and adaptive array processors for vision and other perceptual tasks will be a necessity, a point of confluence with the closely related field of robotics.

Mass production of inscrutable patterns

The knowledge engineer's building blocks are thus *patterns* descriptive of key concepts underlying the given consultant skill. A state-of-the-art system requires many hundreds of such descriptive patterns to be programmed, and the current cycle of development envisages thousands or even tens of thousands for complex task environments. The work of coding even one pattern can consume many programmer weeks, so that the total task appears prohibitive.

Accordingly, the knowledge engineer of the 1980's will not construct his own building blocks, but will have recourse to automated systems of pattern synthesis. Such systems already exist. They must be equipped with stocks of primitive descriptors appropriate to given domains. Pattern synthesis is then induced by supplying tutorial specifications in the form of examples and counter-examples.

Methods have recently been developed which can inductively synthesise patterns from examples for a small fraction of the cost of programming them by hand. When run on the machine in the form of classification programs, machine made patterns typically out-perform man made ones both in accuracy and execution cost. But these machine efficient patterns turn out to be conspicuously different from those developed by experts (Michalski and Chilausky, 1979), and in general to be somewhat inscrutable to humans (Quinlan, personal communication). A methodology is therefore needed for humanising the man-machine channel.

A small number of salient phenomena can now be combined to yield conclusions about the systems towards which we are moving.

1) Knowledge-based systems are memory-intensive rather than processor-intensive. They will soon comprise thousands of stored rules per system.

2) Costs of memory relative to processor costs will continue to decrease.

3) The way to use a large memory as the basis of knowledgeable behaviour is to fill it with *patterns* descriptive of the key concepts of the given knowledge-domain. From these the rule-bases are built.

4) It is becoming possible to mass produce such patterns by machine more cheaply than by programming.

5) The resulting patterns are highly efficient at run-time but their form

tends to inscrutability for the domain-specialist.

6) A preliminary look has indicated that in some cases there may be transformations capable of rendering machine optimised patterns into more humanly transparent forms.

An architecture for knowledge engineering in the 1980s

Putting the above together we can identify major components of future systems. The resulting automated mining-and-refining plant for human knowledge presents some unusual features. The idea will be illustrated with a speculative practical application, chosen for its attractive mix of combinatorial complexity, susceptibility to knowledge-based approaches and commercial potential, namely the identification of organic compounds by industrial chemists.

The skilled chemist performs the knowledge-based computation shown in Figure 16.4. This computation is extremely hard to simulate by program. A decade of work at Stanford University by the DENDRAL project has resulted in a system proficient at identifying straight-chain aliphatic compounds and members of certain classes of oestrogenic steroids. Such expertise is too narrow to be of serious interest to chemists. Some other approach is needed.

As a starting point take one giant look-up memory, conservatively 10^{12} bits†, directly addressable. We wish to use it as a dictionary of mass spectrogram patterns, a likely molecular structure being entered against each pattern. Such dictionaries exist in industrial use but are constructed by hand, and do not exceed 100,000 entries. If the computation is so hard, how can we compute the entries for the dictionary in the first place?

As indicated earlier in this article, we ask: 'What about the inverse computation?' Programs certainly do exist for predicting

in reasonable time. If we could generate exhaustively and irredundantly the complete set of molecular structures in the given class then by

---

† Such a memory can be obtained today for a price of approximately £1M.

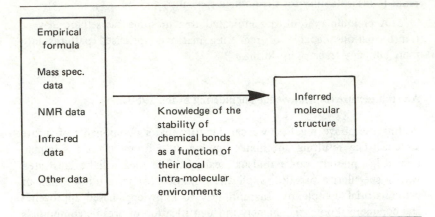

FIGURE 14.4.   The experienced chemical consultant is able to compute the molecular structure of an unknown compound by applying his physico-chemical knowledge, and heuristic rules of thumb, to an assemblage of measurements performed on the unknown compound.

computing for each structure its proper predicted pattern we could construct (back-handedly) the desired dictionary. A suitable structure-generating program exists in the form of Raymond Carhart's (1977) 'CONGEN'.

Such an immensely powerful question-answering resource would unfortunately be limited to answering what I have termed elsewhere 'questions of the first kind' (What is the value of f(x)?) without being able to tell the chemist anything of the 'why'. This is the point at which the AI specialist has to come back into the picture to deploy his inductive inference machinery (software tools such as Michalski's INDUCE and Quinlan's ID3) to compress parts of the dictionary into pattern-rule form. Where possible he must also humanise for intelligibility. We end up, then, with a scenario like that of Figure 14.5. The weakest link in the diagram, because the problem has only recently been identified, let alone solved, is the 'humanisation loop'. Initial study suggests methods applicable in some cases — e.g. for converting large homogeneous decision trees into hierarchically structured collections of human-type rules. In other cases conversion of representation may be thwarted by complexity considerations. Choice may then have to be exercised between sacrificing the superior efficiency of the machine made algorithm, and equipping it with a simplified 'cover story' for human use. The possibility of enabling knowledge-based systems to handle 'stories' in this sense, for purposes

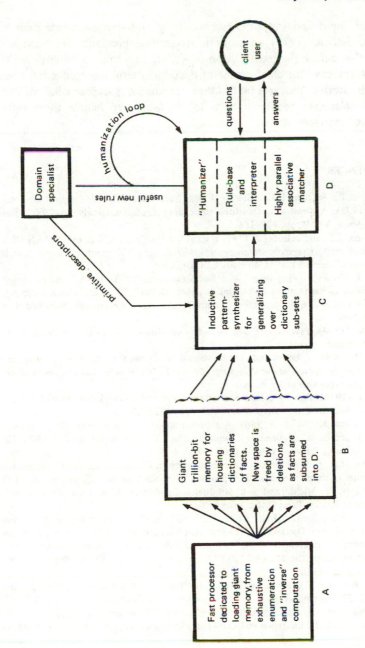

FIGURE 14.5. 1980s scenario for knowledge based computing. Since A and C both represent computations of standard type, they can be combined into a single machine.

both of input and output, deserves study and overlaps work such as that of Schank (1977) on story-understanding programs. We must not forget that before the introduction of writing, and to some extent even to the present day, the chief means of encoding useful knowledge has been through stories, proverbs and other mnemonic paraphernalia of folk science. Machines too may have to be taught to handle these time-honoured representations.

REFERENCES

Ambler, A.P., Barrow, H.G., Brown, C.M., Burstall, R.M. and Popplestone, R.J. (1975). A versatile system for computer-controlled assembly, *Artificial Intelligence*, Vol. 6, pp. 129–156.

Armstrong, J.L. and Jelinek, J. (1977). CHARD User's manual (a FORTRAN CLIP emulator), *Research Memorandum MIP-R-*115. Edinburgh: Machine Intelligence Research Unit, University of Edinburgh.

Bennet, J., Creary, L., Engelmore, R.S. and Melosh, R. (1978). SACON: a knowledge-based consultant for structural analysis, Memo HPP-78-28, also *Report No. STAN-CS-*78-699, Stanford: Computer Science Department, Stanford University.

Binet, A. (1894). *Psychologie des Grands Calculateurs et Jouers d'Echecs*, Paris: Hachette.

Bramer, M. (1980). An optimal pattern-based algorithm for King and Pawn against King, in *Advances in Computer Chess* 2 (ed. M.R.B. Clarke). Edinburgh: Edinburgh University Press.

Bratko, I. (1978). Proving correctness of strategies for the AL1 assertional language, *Information Processing Letters*, Vol. 7, pp. 223–230.

Bratko, I. and Michie, D. (1980). A representation for pattern-knowledge in chess end-games, in *Advances in Computer Chess* 2 (ed. M.R.B. Clarke). Edinburgh: Edinburgh University Press, pp. 31–56.

Buchanan, B.G. (1979). Issues of representation in conveying the scope and limitations of intelligent assistant problems, in *Machine Intelligence* 9 (eds. J.E. Hayes, D. Michie and L.I. Mikulich). Chichester: Ellis Horwood, and New York: Halsted Press (John Wiley).

Buchanan, B.G., Smith, D.H., White, W.C., Gritter, R., Feigenbaum, E.A., Lederberg, J. and Djerassi, C. (1976). Applications of Artificial Intelligence for Chemical Inference, XXII, Automatic rule formation in mass spectrometry by means of the Meta-DENDRAL program, *J. Amer. Chem. Soc.*, Vol. 98, pp. 6168–6178.

Carhart, R.E. (1977). Re-programming DENDRAL, *AISB Quarterly*, Vol. 28, pp. 20–22. This paper gives a brief overview of DENDRAL from a utility standpoint, and also discusses solutions to the structure-generating problem.

Charniak, E. (1977). Inference and knowledge in language comprehension, in *Machine*

Intelligence 8, pp. 541–574 (eds. E.W. Elcock and D. Michie). Chichester: Ellis Horwood; and New York: Halsted Press (John Wiley).

Chase, W.G. and Simon, H.A. (1973). Perception in chess, *Cognitive Psychology*, Vol. 4, pp. 55–81.

Chilausky, R., Jacobsen, B. and Michalski, R.S. (1976). An application of variable-valued logic to inductive learning of plant disease diagnostic rules, *Proc. 6th Ann. Internat. Symp. on Multi-varied Logic*, Utah.

Davis, R. and King, J. (1977). An overview of production systems, in *Machine Intelligence* 8, pp. 300–332 (eds. E.W. Elcock and D. Michie). Chichester: Ellis Horwood and New York: John Wiley.

De Dombal, F.T., Leaper, D.J., Staniland, J.R. *et al.* (1972). Computer-aided diagnosis of acute abdominal pain, *Brit. Med. J.*, Vol. 2, pp. 9–13.

Duda, R., Gaschnig, J. and Hart, P. (1979). Model design in the PROSPECTOR consultant system for mineral exploration, in *Expert Systems in the Microelectronic Age* (ed. D. Michie), Edinburgh: Edinburgh University Press, pp. 153–167.

Fagan, L.M. (1978). Ventilator management: a program to provide on-line consultative advice in the intensive care unit, *Memo HPP-78-16*. Stanford: Computer Science Department, Stanford University.

Feigenbaum, E.A. (1979). Themes and case studies of knowledge engineering, in *Expert Systems in the Micro-electronic Age* (ed. D. Michie), Edinburgh: Edinburgh University Press.

Gregory, R.L. (1970). *The Intelligent Eye*, London: Duckworth.

Groot, A. de (1965). *Thought and Choice in Chess* (ed. G. Baylor) (translation, with additions, of Dutch version of 1946). The Hague and Paris: Mouton.

Hunt, E.B., Marin, J. and Stone, P. (1966). *Experiments in Induction*, New York: Academic Press.

Hunter, I.M.L. (1962). An exceptional talent for calculative thinking, *Brit. J. Psychol.*, Vol. 53, pp. 243–258.

Kunz, J. (1978). A physiological rule-based system for interpreting pulmonary function test results, *Memo HPP* 78-19. Stanford: Department of Computer Science, Stanford University.

Larson, J. and Michalski, R.S. (1977). Inductive inference of VL decision rules, *SIGART Newsletter*, Vol. 63, pp. 38–44.

Martin, N., Friedland, P., King, J. and Stefik, M.J. (1977). Knowledge-based management for experiment planning in molecular genetics, *Memo-HP-77-19*, Stanford: Computer Science Department, Stanford University. Also in *Proc. 5th Inter. Joint Conf. on Artif. Intell., IJCAI-77*, Pittsburgh, Computer Scicence Department, Carnegie Mellon University.

Michalski, R.S. (1978). Pattern recognition as knowledge-guided induction, *UIUC-DCS-R-927*. Urbana: Department of Computer Science, University of Illinois.

Michalski, R.S. and Chilausky, R.L. (1980). Knowledge acquisition by encoding expert rules versus computer induction from examples: a case study involving soybean pathology, *International Journal for Man-Machine Studies*, Vol. 12 no. 1, pp. 63–87.

Michie, D. and Chambers, R.A. (1968). BOXES: an experiment in adaptive control, in *Machine Intelligence* 2, pp. 173–152 (eds. E. Dale and D. Michie). Edinburgh

Edinburgh University Press.

Michie, D. (ed.) (1979). *Expert Systems in the Micro-electronic Age*, Edinburgh: Edinburgh University Press.

Michie, D. (1979a). Machine models of perceptual and intellectual skills, in *Scientific Models and Man. The Herbert Spencer Lectures* 1976, pp. 56–79 (ed. H. Harris). Oxford: Oxford University Press.

Nievergelt, J. (1977). The information content of a chess position, and its implications for the chess-specific knowledge of chess players, *SIGART Newsletter*, Vol. 62, pp. 13–15.

Pople, H.E., Myers, J.D. and Miller, R.A. (1977). DIALOG: a model of diagnostic logic for internal medicine, *Proc. 5th Inter. Joint Conf. on Artif. Intelligence, IJCAI-77*, Pittsburgh: Computer Science Department, Carnegie Mellon University. (This program is now called INTERNIST.)

Quinlan, J.R. (1979). Discovering rules by induction from large collections of examples, In *Expert Systems in the Micro-electronic Age* (ed. D. Michie). Edinburgh: Edinburgh University Press, pp. 168–201.

Schank, R.C. (1977). Representation and understanding of text. In *Machine Intelligence* 8, pp. 575–619, (eds. E.W. Elcock and D. Michie). Chichester: Ellis Horwood; and New York: Halsted Press (John Wiley).

Shortliffe, E.H. (1976). *Computer-Based Medical Consultations: MYCIN*. New York: Elsevier/North Holland.

Waltz, D.L. (1972). Generating semantic descriptions from drawings of scenes with shadows, *MAC AI-TR-271*, MIT, Cambridge, Mass.

Wipke, W.T. (1974) Computer-assisted 3-dimensional synthetic analysis, in *Computer Representation and Manipulation of Chemical Information*, pp. 147–174 (eds. W.T. Wipke, S.R. Heller, R.J. Feldmann and E. Hyde). London and New York: Wiley Interscience.

Zuidema, C. (1974). Chess, how to program the exceptions, *Afdeling informatica IW21/74*. Amsterdam: Mathematisch Centrum.

PART B
Sunday

Introductory Note to Chapters 15, 16, 17, 18, 19 and 20

The 'hand-eye' system mentioned in the first of the following six articles was FREDDY 2, the achievement of a gifted and dedicated team. R.M. Burstall, R.J. Popplestone, H.G. Barrow, A.P. Ambler and C. Brown designed and conducted the programming, and S.H. Salter and G. Crawford were responsible for the design and engineering of mechanical, electronic and optical subsystems. The task specification cited in the first article was that agreed with the sponsors in 1972 as the touchstone of continued support. Against the odds the spec was met, but new advice had meanwhile been received (the 'Lighthill Report' of 1973) questioning the validity of Artificial Intelligence as a basis for work on robotics.

Machine Intelligence in the Cycle Shed appeared in the *New Scientist* on February 23, 1973, the very day that SRC's Board of Visitors was in Edinburgh to decide the future both of the main robot project and of a linked study of computer-based planning using chess end-games. It was later rumoured that at least the chess work would have been funded but for the irritant effect of this article. If there is truth in this, it must amount to the largest negative fee ever received for writing in the *New Scientist*.

During this same epoch the Rothschild Report on Government support of R and D, published as a Green Paper, drew almost universal cries of alarm from the academic world. In those days I saw nothing wrong or strange in the doctrine that even an intellectual person should be asked on occasion to make himself useful. So I set myself to probe the nature of this widely voiced unease. *Pillars of the Tabernacle* contains my conclusions.

I have to confess that having, as I believed, perceived a social evil by no means lacking from my own institution, I thought that to announce my perception might accomplish good. I now see that this, along with other academic activisms, betrayed a misunderstanding of the role which society demands of the University person. Down the ages the academic has been the specialised repository, digester and regurgitator of the precious elixir. On the principle, 'cobbler, stick to thy last', he should not be distracted or harried into processes of large-scale production, marketing or application of the knowledge which he holds in trust. In this he resembles that special caste of bee in some hives whose task is to gorge themselves until they take on the dimensions of honey pots — which in a sense they are. Hanging by their mandibles from the roof they render their huge burdens into nectar, solicited from their mouths by successive troops of their juniors.

Peer Review and Bureaucracy was written after a four-year interval in which I had resolved the theme sufficiently even to feel moved, in *Song and Dance Story*, to offer advice to a fellow scientist. The final piece of the set closes with advice about advice. Looking back, it seems to me now that it is too bland, and that the events which provoked it deserve a final comment.

A historian sees acts of creation and disaster in terms of personalities and consequences. A scientist may perceive here a clearly conceived and superbly manned venture by people of varied backgrounds and ages who saw the possibility of exploring for the first time the epistemological structure of man's oldest domain — the "hand-eye" world. These people were able to combine this with a thrust of relevance to technical needs of manufacturing industry and thus, for a while, to keep uncomprehending critics at bay. Work of excellence by talented young people was stigmatised as bad science and the experiment killed in mid-trajectory. This destruction of a co-operative human mechanism and of the careful craft of many hands is elsewhere described as a mishap. But to speak plainly, it was an outrage. In some later time when the values and methods of science have further expanded, and those of adversary politics have contracted, it will be seen as such. The persons will be long forgotten.

Since those days semi-conductor costs have fallen so low that a new start based on cheap micros has recently become feasible. On present plans an integrated FREDDY 3 will soon exist sufficient for resumption of the experimental study of world models which constituted the earlier long-term goal and raison d'être.

CHAPTER 15
Life with Intelligent Machines

There is a sense in which the study of machine intelligence can be described as the application of philosophy to technology. From mathematical logic insights into processes of reasoning are obtained, and from epistemology ways of representing symbolically what we know about the world. The machine intelligence practitioner is a 'knowledge engineer', who needs to find ways of representing symbolically what his computing system should 'know', and thus equip it to handle the intelligent tasks which he has in mind for it.

Let us consider an intelligent task, that of interpreting mass spectrograms, in order to identify the unknown chemical compounds from which they are obtained. Such a task requires a great deal of sophisticated knowledge about chemistry. E.A. Feigenbaum, J. Lederberg and others at Stanford University, USA, have developed a computer program, DENDRAL, which performs this task. For certain families of organic compounds, identifications are more reliable than those given by experienced human chemists. If we look inside the program we find organised symbolic representations, corresponding to human 'knowledge', of various categories of facts — topological facts about the theoretically possible configurations of atoms to make a molecule of given composition, chemical facts about the stability and non-stability of different configurations, and an entire 'mini-theory' of mass spectrometry. DENDRAL is expected to evolve into a cost-effective tool, able to displace the human chemist over ever-enlarging areas of the mass spectrometry domain. In doing so, it is no more than a forerunner of a 'scientist's aid' type of intelligent computer program which is expected to become commonplace in the years ahead.

Now, consider a different task — that of picking physical components out of a heap, visually identifying them, and using them to assemble a specific object — say, a toy car. This is undoubtedly an intelligent task. It is performed by our Edinburgh 'hand-eye' robot under the control of a versatile assembly computer program in the following stages:

221

Instruction phase

1) Individual parts are tossed onto the platform and the robot is told, for each possible view of the object, its designation, how to pick it up, and what to do with it (e.g., 'turn it over', or 'put it exactly here in preparation for assembly').

Approximately five of these training views are needed for each designation (e.g., 'car body on side', 'car body on back') to assure the system can recognise the part; of course it only needs to be told once what to do with it.

2) Starting with the parts laid out in the fixed position, the robot, working blind, is guided through the assembly operation. The instructions developed at this time to guide the robot constitute the assembly program; thenceforth running the assembly program transforms the laid-out parts into the final product.

Execution phase

1) Someone dumps a pile of parts (perhaps with missing or extra parts) onto the platform and starts the inspection and layout process.

2) The robot examines the platform and lays out any recognised parts for assembly.

3) Any unrecognisable pile of parts is pulled apart into its component parts by a set of routines which can deal with arbitrary heaps of arbitrary parts.

4) If all the parts for assembly are found, extra parts are dumped in a special location. If some parts are missing, the robot appeals for help.

5) The assembly program is run.

The above-described performance is based on an elaborate suite of programs which confer a fair degree of flexibility on the system — it can be taught a new assembly, say a model ship, at a day's notice. How far we still have to go in incorporating 'teachability' into software can be judged from the fact that a three-year-old child can be taught a new assembly in ten minutes! The discovery of better design principles for 'teachable' programming systems is a major goal of most research laboratories in the machine intelligence field.

Over the next decade, what is known as 'software engineering' is expected to develop into one of the world's largest industries. To give an idea of the scale on which the engineering of a single software assignment can be conducted, the writing of one particular program, IBM's OS360 operating system, consumed 3,000 man-years at an estimated cost approaching £30M. Thus we find that here is a new form of manufacture, the operations of which consist in making marks on paper, with a cost so great that one looks for analogies to such enterprises as the pyramid building of the ancient Egyptians. There is a difference, however. One brick of the Great Pyramid can be out of place, yet the pyramid will stand. But one binary digit wrong in a complex computer program can cause the whole to malfunction. Such an error did indeed cause failure of one of the unmanned missions of NASA's space program. In software, there is a premium on accuracy which is quite unprecedented.

Techniques of machine intelligence can come in at this point *via* an unconventional interpretation of the principle of redundancy. Classical redundancy operates by replication of *similar* elements. Thus, if a signal is regarded as verified if, and only if, inputs from three elements all agree, and each of these has an independent error rate of 1%, then the error of the transmitted signal is automatically reduced to one in a million. In a machine intelligence program, as in the case of versatile assembly instanced above, the fundamental design principle is 'if at first you don't succeed, try something else!' For a given task — say picking a component out of a heap — selection is made from a battery of *dissimilar* strategies — 'grab a protuberant part', 'seek isolated object and identify', 'stir heap and start again', etc. The program is quick to abandon any given strategy if un-rewarded by success, and to go back to the strategy-pool. The result is an impressive reliability of eventual goal-accomplishment: sooner or later the program is bound to win through to the given objective. This kind of reliability has not in the past been associated with the behaviour of auto-matic systems. Rather, opportunistic flitting between alternative strategies in the light of what turns up is more characteristic of the way that a human tackles a skilled task — whether in driving a car, cooking a meal, or proving a theorem.

Three years ago I discussed the question of 'ripeness of time' with regard to the project to construct an intelligent machine. Although my tentative conclusions were positive, there was nothing then stirring in the world to give them particularly forceful backing. But, less than one year later, in July 1971, the Japanese Government announced their 'PIPS project'. PIPS stands for 'Pattern Information Processing System' and its goals have been characterised in the following terms by an expert American observer, Dr Gilbert Devey of the US National Science Foundation. '. . . an inanimate system capable of sensing a pattern (characters, pictures, 3-D objects), identifying the nature of that pattern, relating that information to a data base of information (scene analysis) to then decide on and take a course of action for the control and manipulation of an output device which performs a useful function, and doing so without human intervention'.

This unprecedented R and D project is to a major extent concentrated on what I refer to as 'Meccano problems'. Whether we like it or not, an internationally competitive program to develop intelligent robots has begun. The United States Government, already investing substantially through its Department of Defence, is responding to the Japanese challenge through additional agencies, notably NASA and the National Science Foundation. Independently of this, activity in robotics research is gathering force in numerous American laboratories, for example, the Draper Laboratories in Cambridge, Massachusetts, and IBM's Thomas J. Watson Research Laboratories at Yorktown Heights. Let us consider what possible consequences to our culture and civilisation may ultimately flow from the present world-wide intensification of effort. Whether or not we in Britain decide to go with the wave, it is now unlikely to halt.

Where will it ultimately carry us? Are we, or our children, or our grand-children to share this planet with an alien race of equal, or possibly superior intelligences? If so, is this good or bad? Will new windows of the mind be opened to us, or will we face the ultimate in environmental pollution — pollution with uncomprehended machinery?

An indication that educated people believe that 'it is later than you think' with regard to these questions is given by two recent surveys of opinion which I.J. Good and I made independently of each other. Good (1973) writes:

You asked me to send statistics of the estimates of when a machine would be built with the intelligence of a man. I put the question in the following manner. People were invited to say by which date they thought there was a half chance that the machine would be built, among the eight dates listed below. The answer was to include the belief that it is odds against an intelligent machine's ever being achieved. I asked the question on two occasions. The first occasion was at an after-dinner speech in March at a meeting of Virginia computer users, about half of whom were college men and half were business men . . . The second occasion was at my lecture at the Institute of Contemporary Arts in April. The results were as follows:

		1980	*1990*	*2000*	*2020*	*2050*	*3000*	*5000*	∞
Virginia	(i)	0	6	20	16	5	3	0	4
ICA	(ii)	0	8	24	12	5	3	0	16

My own questionnaire was circulated among scientists of several different laboratories in America and Britain working in machine intelligence and in closely related branches of computer science. I reproduce the questions in Table 15.1, and have entered the frequencies with which the various replies were given.

TABLE 15.1

Aggregate results of a questionnaire completed by sixty-seven computer scientists working in machine intelligence and related fields (Spring 1972).

1) It will have become a technical possibility to construct a computing system exhibiting an all-round intellignce approximating that of adult humans in 5 years *0* 10 years, *1*, 20 years *16*, 50 years *19*, more than 50 years *25*, never *3*, no opinion *3*.

2) Significant industrial spin-off from machine intelligence research can be expected in 5 years *30*, 10 years *28*, 20 years *4*, 50 years *1*, more than 50 years *2*, never *1*, no opinion *1*.

3) Significant contributions to studies of the brain are likely to come from machine intelligence in 5 years *15*, 10 years *20*, 20 years *9*, 50 years *4*, more than 50 years *7*, never *3*, no opinion *9*.

4) Significant contributions to machine intelligence are likely to come from studies of the brain in 5 years *1*, 10 years *14*, 20 years *19*, 50 years *4*, more than 50 years *5*, never *10*, no opinion *13*.

5) 'Machine intelligence' (still less the study of intelligence more generally) is not a unitary discipline and is destined to be overgrown by its constituent disciplines: agree *22*, disagree *35*, no opinion *10*.

6) If the goals of machine intelligence are realised, the immediate effect will be that human intellectual and culture processes will: atrophy *1*, be enhanced *50*, be un-affected *9*, no opinion *7*.

7) The risk of an ultimate 'take-over' of human affairs by intelligent machines is: negligible *30*, substantial *26*, overwhelming *5*, no opinion *6*.

Among the points of significance are:

1) that a majority believe that human intelligence levels will be reached in fifty years or less. Other surveys have given even more optimistic estimates, if that is the right word.

2) that about half the sample believed that the risk of ultimate take-over is at least substantial.

These results could be used to justify unease, expressed by Arthur Clarke in a recent exchange with Good. The latter remarked that 'If we build an ultraintelligent machine, we will be playing with fire. We have played with fire before, and it helped keep the animals at bay.' Clarke's reply was 'Well, yes — but when the ultraintelligent machine arrives *we* may be the "other animals"; and look what's happened to them'.

I shall suggest that such prognostications are certainly premature and probably without foundation. It is not that I personally doubt the technical *possibility* within the time-scales here envisaged of constructing an all-round mechanical intelligence of human or superhuman intellectual power. But I doubt whether human motivation could ever exist for implementing the machine equivalent of an intellectual all-rounder. Under stable conditions, it is not *undifferentiated potential* which the employer of labour seeks to hire, but *specialist skill* relevant to his current need. Why, one wonders, should the matter stand differently with regard to machines?

As far as a trend can be observed in computer technology, it is a trend towards speciation. Task-specific capability becomes increasingly incorporated into fundamental system design. Scratch-pad calculations for engineering and statistics are moving from the time-shared terminal to the desk-top computer; large-scale number-crunching is being passed to specialised giants such as the CDC7600, and heavy-duty commercial record-handling to the standard IBM ranges. Suspending disbelief for a moment in Good's Ultraintelligent Machine (Good 1965), can we extrapolate this trend, and envisage populations of machines, each capable within its own narrow repertoire of vast and superhuman powers, but helpless otherwise? The concept is reminiscent of H.G. Well's vision of functional specialisation among the Selenites, the inhabitants of the moon.

The moon is indeed a kind of super-anthill. But in place of the five distinctive types, the worker, soldier, winged male, queen and slave of the ant-world, there are amongst the moon-folk not only hundreds of differentiations, but, within each, and linking one to the other, a whole series of fine gradations. And these Selenites are not only merely colossally superior to ants, but, according to Cavor, colossally, in intelligence morality and social wisdom higher than man (Wells 1956: 217–18).

He describes co-operation among lunar being Phi-oo, Tsi-puff and two other Selenites, in the task of learning Cavor's language. Wells's account is strongly suggestive of the integrated action of three specialised computers: a language processor, a database machine, and a display device:

Phi-oo would attend to Cavor for a space, then point also and say the word he had heard. The first word he mastered was 'man', and the second 'mooney' – which Cavor on the spur of the moment seems to have used instead of 'Selenite' for the moon race. As soon as Phi-oo was assured of the meaning of a word he repeated it to Tsi-puff, who remembered it infallibly . . . Subsequently it seems they brought an artist with them to assist the work of explanation with sketches and diagrams – Cavor's drawings being rather crude. 'He was', says Cavor, 'a being with an active arm and an arresting eye', and he seemed to draw with incredible swiftness (Wells 1956: 217).

The analogy with interaction between a suite of special-purpose computing devices is irresistible. Well's narrative continues '. . . some adjectives were easy, but when it came to abstract nouns, to prepositions, and the sort of hackneyed figures of speech by means of which so much is expressed on earth, it was like diving in cork jackets. Indeed, these difficulties were insurmountable until to the sixth lesson came a fourth assistant, a being with a huge, football-shaped head, whose forte was clearly the pursuit of intricate analogy' (Wells 1956: 17–18).

So might a human scientist of the future, wrestling with a problem of conceptualisation, call to his aid some remote and wonderfully elaborated descendant of Pat Winston's computer program for forming analogies which gained its author well-merited visibility at the Artificial Intelligence Laboratory at MIT.

Here are some questions we must put if we have serious anxiety that intelligent machinery will constitute a threat to man's ego and identity. Should Phi-oo be jealous of Tsi-puff, because of his wonderful memory? Should Tsi-puff be consumed with envy of Phi-oo's linguistic excellence? Should they both feel threatened by the third Selenite's draftmanship or by the fourth Selenite's grasp of analogy? Surely not. Man is a territorial animal, and his patterns of rivalry and co-operation extend to intellectual territory too. Where skills are complementary and competition pointless, the sense of trepass does not ordinarily arise. When a man has a specialist skill which machine advances render obsolete he may indeed feel resentment. Professor Aitken was the world's most outstanding calculating prodigy and is reported to have felt concern at the growing numerical powers of computers. This distress occurs and will continue to occur, and

is part of the general social problem of mechanisation. But in each generation the shoe pinches in a different place. Perhaps the only *specialist* skill
which human intellectual workers will develop in the world of Good's
'ultraintelligent machine' is that of managing and co-ordinating teams of
intelligent machines.

Such machine systems may require rather careful and wise management
as time goes on, if only to prevent their human beneficiaries becoming
helplessly dependent upon them — parasites instead of symbionts. The
reality of the problem is underlined by the news, announced in June 1972
by Y. Masuda, director of the Computer Usage Development Institute,
that the Japanese Government's £2500M plan for computerisation includes the development of a prototype 'computer-controlled city' within
the next ten years.

However, such computer-controlled cities belong to the future, and
autonomous 'Mars rover' vehicles collecting our garbage or tidying up
warehouses, or handling airport luggage will not affect our lives much, for
all their science fiction aspect. What *will* begin to colour our existence by
the end of this decade is the emergence of the 'home terminal' offering
an intelligent 'question-answering' facility as a service.

Not only schools, hospitals, and commercial firms, but also the
ordinary householder will be able to tap information and problem-solving
power from a national computing grid, rather as he now draws on gas,
water, and electricity. Computer-aided self-instruction will have become a
hobby of large sectors of the population by the turn of this century.

When home terminals can offer a useful service, the citizen will cease
to regard the computer as a monster or a competitor. Instead, the conversational terminal of the future will be welcomed for what it will do to enlarge daily life — as planning assistant, as budgeting assistant, and above all
as a knowledgeable and challenging tutor and companion. When I look
ahead towards the coming symbiosis, I naturally see negative aspects, and
dangers which will have to be carefully watched. But the main impression
is of a more varied and stimulating world, with prospects of man being
culturally and intellectually master in his own house as never before.

REFERENCES

Clarke, A.C. (1968). The Mind of the Machine. *Playboy*, December.

Devey, G. (1972). Project Management for Automation Technology. A paper delivered at the Japan Industrial Technology Association (JITA) Symposium on Pattern Information Processing Systems. Tokyo, March 15.

Good, I.J. (1965). Speculations Concerning the First Ultraintelligent Machine. *Advances in Computers* 6: 31–88. New York: Academic Press.

Good, I.J. (1973). Personal communication.

Wells, H.G. (1956). *The First Men in the Moon*. London: Fontana Books.

CHAPTER 16

Machine Intelligence in the Cycle Shed

During the 1970s, computing in its various forms is expected to become the world's third largest industry, with the software business predominating. Thereafter, the development of self-programming systems exhibiting some degree of 'intelligence' promises ultimately to transform our whole economic and cultural life. But these opportunities will not come unbidden.

Britain is both well qualified and ill qualified to play a leading part here: well qualified because of its rich concentration of talent in advanced computer science; ill qualified because most of the sites of concentration are in the enchanted playgrounds of the universities, where criteria of unstructured excellence are paramount and cost-benefit considerations are not always felt to be in the best of taste.

It is no solution to seek to re-locate these brain-banks in industry, where the pay-back horizon is of the order of three years. In advanced computing, 'strategic research' — with a horizon of 10 to 15 years — plays a dominant role. And in any case, some aspects of the 'playground' ambience of the university constitute a real aid to creativeness. The problem of providing appropriate and stable conditions for this kind of work is receiving attention from the Research Councils as part of the post-Rothschild debate. One can trust with fair confidence that the right administrative framework will be engineered eventually. But I want here to consider some of the more intangible difficulties.

Improvement of contemporary programming systems in the direction of 'learning' and 'problem-solving' represents a deliberate encroachment on the functions of the human intellect. To some, this conveys an aura of the exotic, the audacious, even blasphemy. The research seems hard to relate to any existing field. Difficulties of placing novel fields, and indeed individual discoveries, are by no means new or uncommon; they are probably related to the notion of prematurity. In his perusals of the scientific literature and consultations with the grape-vine, every scientist comes across the phenomenon of premature discovery — meaning the acquisition of new knowledge which is destined for ultimate recognition, but which is so far ahead of its time as to be overlooked or dismissed. Most scientists regard such mishaps as regrettable but not significant. After all, if X didn't discover it or obtain recognition for it, then Y and/or Z would have anyway.

This phenomenon was re-analysed in last December's issue of Scientific American, by Gunther Stent, professor of molecular biology at Berkeley. Stent related scientific innovation to innovation in art and literature. He contended that new work, whether in science or in art, can gain acceptance only if it can be connected to contemporary 'canonical' knowledge. Unless, and until, such connections can be constructed, a scientist's discovery (such as Mendel's 1867 paper, rediscovered in the early 1900s) or an artist's creation (such as Picasso's first cubist painting 'Les Demoiselles d'Avignon', done in 1907) will remain unappreciated. This is not because the professional world is wicked or foolish, but because the innovator's product — though it may be sound or even brilliant — is in a rather literal sense useless.

Stent also considers the possibility that an entire field of enquiry may lack connection not with its own canonical body of knowledge (this would be a contradiction) but with the body of knowledge possessed by the scientific community at large. This is precisely the situation of machine intelligence research today, where techniques of computing are used to investigate questions not of natural science but of philosophy. The fact that the search for formal (and hence mechanisable) definitions of phenomena of thought and knowledge may yield technological rewards is irrelevant to the present argument: the established specialism to which machine intelligence most logically belongs is too remote from practical science to facilitate general acceptance.

Has this done any particular harm so far? The following circumstances seem relevant.

1) Advances in nuclear physics (with the possible exception of plasma physics) are not expected to be of industrial, military or social significance.

2) The Science Research Council's budget for nuclear physics runs at about £20 million per annum.

3) The SRC's expenditure on research in computer science is less than

£1 million per year (of which perhaps a tenth is devoted to machine intelligence).

4) Computing is the example par excellence of a brains-intensive industry; whatever hopes Britain has of economic revival must lie in such areas.

By contrast, when a technology has a well-established canon into which novel concepts can be absorbed, eagerness to invest may outweigh all considerations of a project's expensiveness and likely unprofitability. (An unkind observer of Concorde's present plight might propose it as a case in point.)

	Inexpensive	Expensive
Economically profitable	A	B
Economically unprofitable	C	D

Contemplate the diagram above. A reasonable man might expect national priorities to be in the order A, B, C, D. But there is a peculiarly British mannerism of government, whereby a virile ability to spend £500 million is matched by maidenly prudishness about pennies. Northcote Parkinson has decribed a related phenomenon — the time taken to reach a decision being in inverse proportion to the sum involved. He illustrates this with an imaginary committee meeting:

Allowing a few seconds for rustling papers and unrolling diagrams, the time spent on Item Nine will have been just two minutes and a half. The meeting is going well. But some members feel uneasy about Item Nine. They wonder inwardly whether they have really been pulling their weight. It is too late to query that reactor scheme, but they would like to demonstrate, before the meeting ends, that they are alive to all that is going on.

Chairman: 'Item Ten. Bicycle shed for the use of the clerical staff. An estimate has been received from Messrs Bodger & Woodworm, who undertake to complete the work for the sum of £350. Plans and specification are before you, gentlemen.'

Mr Softleigh: 'Surely, Mr Chairman, this sum is excessive. I note that the roof is to be of aluminium. Would not asbestos be cheaper?'

Mr Holdfast: 'I agree with Mr Softleigh about the cost, but the roof should, in my opinion be of galvanised iron. I incline to think that the shed could be built for £300.'

Mr Daring: 'I would go further, Mr Chairman. I question whether this shed is really necessary. We do too much for our staff as it is. They are never satisfied, that is the trouble. They will be wanting garages next . . . '

And so on. I am not suggesting that machine intelligence research is, like the bicycle shed, of trivial importance; quite the contrary. What it shares with the shed is that it is of trivial cost when measured on an appropriate scale — relative to, say, nuclear physics. If the world of government science were like the world which Parkinson delights to caricature, we could imagine discussion of Item Ten spinning out interminably — decision is postponed from meeting to meeting; a special panel of bicyclists studies the shed-building proposal for a year, and then reports positively; an eminent Cambridge physicist is called in and pronounces the whole concept to be unsound; eventually the feeling takes root that 'shed' has become a four-letter word and that the wise man should steer clear.

Happily we are not imprisoned in Parkinson's world. Returning to the 'A, B, C, D' table we should confidently (1) assign advanced computing science, including machine intelligence an 'A' for top priority, (2) terminate discussion of Britain's bicycle shed, and (3) get on with it.

CHAPTER 17
Pillars of the Tabernacle

The instinct of self-preservation normally restrains a man, especially if he is an academic, from publicly questioning his own right to exist. What are we to say, then, when an entire profession rises up, as academic science did in reaction to Rothschild, and self-righteously claims to be beyond reach of justification?

Rothschild said in effect that if a scientist is to be maintained from public funds then some return to the tax-payer should be visible. Returns can be of two kinds, cultural and technological. But the academic scientist typically declares himself inaccessible to normal procedures of justification, on the grounds that the cultural benefits of academic science are intangible and the ultimate consequences for technology not calculable. *Contributions to culture:* Advances in cosmology, or in the interpretation of human fossil remains, or in the theory of numbers, or in the decipherment of Hittite scripts, or in the hunting of the quark, belong with the creation of new symphonies, paintings, poems, novels or theologies. The citizen is prepared to pay for a certain amount of this. *Contributions to technology* embrace whatever seems to the administrators of science to be sufficiently quantifiable to permit of investment calculations. Thus medical research has entered an era in which accountants can count costs and economists assess savings. The industrial burden imposed by the common cold has been intensively studied by economists (and rightly studied) in advance of any significant therapeutic discoveries. With lung cancer the calculation is sharpened by the need to include the shortening of human life on one side of the balance-sheet, not to mention (on the other side) the economic value of the tobacco companies.

The citizen is prepared to pay for an indefinite amount of R and D, whenever he is persuaded that someone has done the sums and has shown that there is a reasonable expectation of social profit. Unscrupulous or self-deceiving persons can, and do, exploit this willingness by 'selling' costly *cultural* ventures as technological. Concorde may have been motivated by aesthetics, love of adventure, national pride, prestige, rivalry, curiosity, or any combination of these with other abstract themes: even the quest for immortality, as with the building of the Pyramids, may have come into it. The one motive which seems excludable has been the pursuit of economic return.

I do not, by the way, imply a criticism of the ancient Pharaohs, for they were at least straight-forward about what they were doing. They did not recruit their labour force and exact their finances on the pretext that the Pyramids would be useful as sundials.

Concorde and the Pyramids were marvels of technology in the service of culture.

Let us return to academic science.

It is anybody's game to contest, refine, revise, refute or endorse any given exercise in the accountancy of scientific research. What does *not* seem open to doubt is the fundamental principle: that in conditions of economic stringency some accounting must be made. A cry of grief at once arises, provoked by a sense of loss for something precious. Belligerent postures are adopted, as though to defend the approaches to a sanctuary. It is, indeed, just so. There *is* a threatened sanctuary, and it conceals Academe's most secret treasures. Those who live and work in its shade know that our Tabernacle is builded upon one great freedom — *the freedom to act as trivially as we like*. At the four corners stand four supporting freedoms. They are:

1) *First Supporting Freedom.* If I have a thought, who can measure it, what can contain it? Am I not king of infinite space? One thought, however vacuous, outweighs a world of actions, or so I am free to believe.

2) *Second Supporting Freedom.* I am free to insist that my pure pursuits *could* have immense social, industrial, medical, agricultural or military potential, if only I could be bothered to make the relevant theoretical or practical demonstrations.

3) *Third Supporting Freedom.* I am free to stigmatise as a tradesman anyone caught on campus in the act of such demonstrations.

4) *Fourth Supporting Freedom.* I am free to cease from all identifiable activity, while continuing to draw my pay.

It would be unfair to accuse the general run of academic scientists of taking undue advantage of these freedoms. As with privileges and guarantees in general the value to their possessor chiefly lies in the subjective knowledge that they exist. We must also recognise that the stages in a scientist's maturation at which the different freedoms first reveal their attractions to him are widely spaced out. We would not go far wrong in identifying the age of 20 for Supporting Freedom 1, the age of 30 for Supporting Freedom 2, the age of 40 for Supporting Freedom 3 and in conceding that it is not until the attainment of his half-century that the University scientist feels ready to add the Fourth Supporting Freedom to his Pillars of the Tabernacle.

There is, it should be added, one more Pillar. It rises heavenwards from the roof, and is in fact a flagpole. It offers the freedom to show independence of mind by the emblazonment of innocent lampoons against the Tabernacle — innocent, that is, of the intent to reform. I ask indulgence now if I put out more flags:

1) Rothschild in a recent speech warned of impending national decline, and called for scientific brain-power to be mobilised in a new Dunkirk spirit. This text should be distributed as required reading to all science graduate students in British Universities.

2) The 'task force' idea promoted at different times by Peter Kapitsa and by Hermann Bondi should be revived. At any given moment there are scientific problems of special timeliness, in which a concerted effort employing a diversity of skills could yield predictable returns. Scientists could be freed for such projects by secondment from their employing institutions. As long as a project lasts the task force would enjoy budgetary flexibility and immunity from administrative harrassments. Such opportunities can inspire scientists, as happened during the 1939–45 war, to outstanding feats of innovation.

Alternatively our British destiny may be to serve as a tourist playground, boldly placing our cultural heritage in the shop window, our academic scientists in the Tabernacle and our technology on the shelf. The choice is ours and no-one else's.

CHAPTER 18
Peer Review and the Bureaucracy

Proposals for the reform of refereeing systems have recently been aired. A forgotten aspect is that 'peer review', as far as research funding is concerned, is operated not by the scientific peers themselves but by the officials who look after the advisory committees of the grant-giving bodies.

Usually (but not necessarily) in consultation with the committee chairman, it is they who select the referees for a proposal. It is thus not possible to do justice to 'peer review' without examining the relations prevailing between scientists and bureaucrats. If we can understand how the two worlds meet, or fail to meet, ways may suggest themselves for improving the process.

Last October (1977) *Scientific American* had an article on the United States National Science Foundation's peer review system. An elaborate statistical survey, designed to pick up possible sources of bias (such as 'eminent' applicants being favoured, or applicants from high-prestige universities) led to the conclusion that no important biases are present and that the workings of the system are pretty fair.

This should surprise no one. Anyone who has served on, say, a research council advisory committee can see for himself that in 95% of the cases (perhaps more) assessments are made as objectively as one would wish to see. In the remaining few per cent, however, someone or something is involved to which the 'office' is specifically sensitive. Then matters are handled (not necessarily on the surface) in a different way. Naturally these exceptional cases are swamped in the statistics.

A fairly senior official of a non-British agency, whose nationality I shall not disclose, once told me that if for any reason he felt justified in short-circuiting the system in order to get a given result, he would make a judicious selection of referees — either the scientist's particular friends or his particular enemies, according to which result he wanted. He needn't have told me. I knew it already. In his heart of hearts so does any scientist who has been in the game any length of time.

I do not myself think that anything much can be done to stop this. It is part of the human social condition in its possibly unedifying but very real Monkey Hill aspect. A grown person can reasonably be expected to cope. A scientist can face his critics down, he can ask to see referees' reports (by rightful entitlements in America), or he can follow up with a better drafted proposal. If he's convinced that he's not getting honest judgements he can force even ill-disposed or jealous scientific rivals into partial concession by technical confrontation. He can also, of course, and this is not absolutely unknown, locate the blocking colleagues and treat them to a little discreet stroking, trading and squaring. Admittedly in so doing he dips the flag of science to Monkey Hill.

Unfortunately, when thwarted in getting 'the right answer' from peer review, officials do not always do what they should — disclose that because of additional, non-scientific, considerations the proposal has anyway to be rejected, or accepted, as the case may be. Being too bashful to do this (possibly someone higher up will be out of sorts if 'the wrong answer' prevails) the office may press on earnestly into various finesses. At this juncture the scientific bystander may have to choose whether to look the other way, or to rock the boat.

The inhabitants of the two worlds inevitably and inescapably differ in what they respectively hold sacred. A scientist gives terrible offence if he tears a civil servant down as he would a seminar speaker who was glossing over something in his technical presentation. An official (but not a scientist) who is made to contradict himself publicly, or is confronted with an unexpected document, or is accused circumstantially of misjudgment, feels his personal sanctuary to be desecrated.

Yet the coin has two sides. The civil servant possesses little awareness of where lies the scientist's holy ground. The official is often needlessly scrupulous to avoid implying disparagement in the personal realm. Many a scientist does not care whether he is *personally* disparaged or not (scientific disparagement from his own peers would be different). Yet while being scrupulous where he need not be, the official may not think twice, if policy seems to demand it, about intruding his people-oriented methodology into issues of scientific truth and value.

Of course scientists need not feel any compulsion to accept the bureaucrat's view of science. But they should probably forgive those who do. Ibsen's play *Brand* offers a wonderful parable. The man of vision and principle, the village priest — how destructive and tiresome his vanity and single-mindedness proved to be for the villagers! The venal, pragmatic mayor, full of tricks and frailties, is the one in whom one recognises the power to conserve, heal and give aid when needed.

What can be done then to approximate North and South? One simple measure would be to require possession of a science PhD as a condition of employment as an officer of any scientific agency of Government. There is no way to soak up the ritual attitudes and moral lore of science except by working for several years as an organic part of a scientific laboratory, as every graduate student does.

The reverse of course, is that after he has satisfied the university's examiners, the applicant for a higher degree should perhaps have his award deferred until he has spent a year working as a dogsbody in Whitehall and can supply a letter of commendation from those he has devilled for.

This simple scheme has another attraction. There is much concern about the overproduction of science PhDs and about their employability. On the new basis the market would be expanded to include a niche for graduate students ready to be coaxed from the laboratory into careers in science administration.

CHAPTER 19
Song and Dance Story

A cry of pain has arrived on my desk — an article by an Australian university zoologist entitled 'Peer review: a case history from the Australian Research Grants Committee' (*Search*, vol. 10 p. 81). The author, Dr Clyde Manwell, relates a tale of unhappy outcomes, first of a series of applications he made to the ARGC for research support, and secondly a number of energetic subsequent attempts to discover why these particular applications had been rejected.

The quest for explanations eventually involved his head of department, a former vice-chancellor, another vice-chancellor, and a member of parliament, all of whom helped procure for him replies which he had otherwise been unable to obtain.

At one point in his article Dr Manwell quotes from my essay 'Peer review and the bureaucracy' (*Times Higher Education Supplement*, 4 August, 1978). This essay discussed, among other things, a recent statistical survey by three American authors seeking to discover whether the peer review system of allocating research money suffers significant biases. The survey's authors concluded that there was 'no evidence to substantiate recent public criticism'. My *THES* article made the point that provided the proportion of such cases were small (say 5% or less of all applications) quite gross abuses could occur and yet they would be swallowed from sight in the statistics. As Dr Manwell points out, other inadequacies of the American survey have also been noted. So the question of peer review — its purity, utility, accuracy and effectiveness, not to mention its costs — remains open.

When I began to read Dr Manwell's article I was expecting him to analyse these more general aspects of peer review, and perhaps come up with proposals for improvement. As I continued, I realised that someone suffering from a sense of injustice, personal bruising and bewildered despair cannot reasonably be expected to play the elder statesman. So what happened to Dr Manwell, exactly?

240

Almost anyone familiar with the ways of the world will conclude that details concerning the arbitrary termination of his grant in 1971, followed by unsuccessful proposals for renewed support in 1971, 1972 and 1976, are almost irrelevant compared with one solitary paragraph near the end which reveals that in 1971 Dr Manwell

... criticised the SA Department of Agriculture's fruit-fly spraying programme. Within a few weeks of those criticisms (repeated in the following years by other scientists with ultimately some improvement in the programme) an attempt was made to dismiss the author from his position. It required a four-year fight to have the author's name cleared of charges laid by a very eminent Australian scientist, which are now officially recognised as 'a number of errors' (Vice-Chancellor's statement 3 June 1975, available from the Registrar, University of Adelaide, SA, 5001).

Manwell adds that it was shortly after the sacking attempt that his grant was terminated without stated reason.

Let us take the hypothesis that a combination of the matter and the manner of his criticisms had indeed displeased some powerful insider. On such a supposition, what should Dr Manwell then have done? Clearly something different from the 'valiant for truth' campaign on which he doggedly embarked. But what?

After some 30 years in the world of science (which is not as different from the rest of the world as is sometimes thought), I would say the following:

First whatever you do, don't become more attached to your wrongs than to your rights.

Secondly, purge the mind of illusion. Peer review, and the rest of the apparatus of grant administration, is not a *court of justice*. It is an *arena*. Points can certainly be scored by demanding the appearance of equitableness. But that is not even half the battle.

Thirdly, ask yourself the question: 'Once an influential insider has been fundamentally antagonised, what remains for me?' Options are (a) tough it out (as Dr Manwell is doing) with or without litigious pinpricks to the persecuting body; (b) raise an army of even more influential insiders for a decisive counter-attack; (c) patch it up (without crawling); (d) crawl; (e) emigrate; (f) leave science and take up investigative journalism.

If the idea of being a plucky loser has an appeal, or if the possibility of achieving an ounce of institutional reform outweighs a ton of damage to your scientific work, plus loss of time, sleep, faith, hope and charity, then by all means consider option a. Having considered it, consider dropping it.

Option b is not, in normal circumstances, even remotely feasible. It takes an awful lot to beat City Hall.

Option c is of course ideal. But in most cases it requires both diplomatic flair and insider status. Dr Manwell in all probability lacks at least one of these.

Option d requires knowing the limits of your own crawl-capacities. These may be more elastic than you think.

If professional opportunities and personal constraints permit option e, then take it, and don't ever look back.

As for f, if Dr Clyde Manwell is half the scientist that I take him for this option will not even have crossed his mind. When the time comes, though, he might consider as a retirement project preparing a readable and well documented monograph on officialdom in science.

'The future is a place' (wrote the poet Mayakovsky)
'In which officials disappear
And in which there is much
Dancing and singing!'

CHAPTER 20
Scientific Advice to Governments

The Lords' Select Committee on Science and Technology is turning its attention to the ways in which the government gets its scientific advice. Ancient Greece and Victorian England knew something of how it should be done. Even before the onrush of technology governments always turned, in the last ditch, to technical advisers of one sort or another – in ancient times to oracles. The qualities looked for have varied. The Athenians approached the Delphic oracle in the 5th century BC when faced with the Persian threat. Its advice was ambiguous:

> Zeus grants the Triton-born a wooden fort
> To stand unharmed and be a last resort . . .

Was the 'wooden fort' the city's Acropolis or the Athenian navy? Themistocles, who had already talked his colleagues into parting with the cash for 200 ships, argued ingeniously for the latter interpretation, and was conspicuously vindicated by the naval victory at Salamis.

In the 4th century BC Dionysius the Younger, ruler of Syracuse, had begun to supplement his sources with a new kind of adviser. Plato had commended abstract thinkers as the best of all possible influences on government. As a politically minded mathematician, he himself typified this new breed. Dionysius's father had known Plato, but after hearing him lecture decided not to listen to him further, and reverted to the traditional oracles. Legend has it that the older Dionysius was so affronted by the lecturer's tedious praise of abstract justice, for which as a tyrant he found little occasion, that he had Plato seized and sold as a galley-slave, While conceding the reality of the 'adviser nuisance', modern taste may discern here an element of over-reaction.

The circumstances of Dionysius the Elder's death are ironical in this connexion. He was told by the oracle:

> You shall not die until you have defeated your betters.

Being in the middle of a war against Carthage he superstitiously held off from decisive military engagement. In 367 BC, however, a play of his own composition won first prize at the pan-Hellenic drama festival, the *Lenaia*. His death followed from excessive celebrations on receipt of the news.

Plato's idea was that such a man as himself, the first-ever 'academic', was especially valuable in government. If he would not, or could not, be a ruler then at least he could advise rulers. He exemplified that kind of intellectual excellence which sets mathematical abstraction on a higher spiritual plane than the concerns of work-a-day mankind, and is evidently well equipped to attract the admiration of administrators. In his 7th letter he writes:

I came to the conclusion that the condition of all existing states is bad — nothing can cure their constitutions but a miraculous reform assisted by good luck — and I was driven to assert, in praise of true philosophy, that nothing else can enable one to see what is right for states and for individuals, and that the troubles of mankind will never cease until either true and genuine philosophers attain political power or the rulers of states by some dispensation of providence become genuine philosophers.

Unfortunately 'academic man' sometimes combines with his special gifts the generalised one of being conspicuously wrong on ordinary questions of the day. On such occasions it is as if his judgement were usurped not by random conjecture but by a cognitive demon able to construct reliably *false* conclusions, provided only that the matter lies outside the pencil-beam of his specialised training. N. Parkinson describes a committee adviser with this aptitude, and introduces the theme with the question —

Failing a man who is always right, what if the organisation contains a man who is always wrong? . . . Why not ask him and then do the opposite?

Whether or not the judgement of academic scientists is indeed capricious beyond the norm, although widely assumed, is not entirely clear. A counter-example can be cited in one of the greatest of all academics, Isaac Newton. While occupying the Lucasian Chair of Applied Mathematics in the University of Cambridge he proved himself a wise and capable public man, negotiating on behalf of his university to good effect, and subsequently overseeing the Mint. Against this stands the advisory career of a later holder of that chair, the 19th century Astronomer Royal, Sir George Biddel Airy. His research achievements continued at a high standard into extreme old age. But his advice to the government was uniformly, and sometimes preposterously, wrong. He stated that if the Royal Salute were fired outside the Crystal Palace the building would collapse. His advice on public funding of research in advanced computation, i.e. of Charles Babbage's difference engine, was more subtly wrong and is now believed to have had a complex motivation. Babbage had applied for a copy of the astronomical observations kept at the Royal Observatory, then directed by Airy. His request was refused, and on investigation he discovered that

Airy was in the habit of selling the newly printed *Greenwich Tables* in bulk as waste paper, and had already disposed of more than five tons to one purchaser. Babbage commented, undiplomatically, that this seemed an extravagant mechanism for giving supplementary remuneration to a public man. The extravagance was however, probably less than the ultimate cost to the nation of Airy's counter-attack. On his advice, government support for Babbage's difference engine was withdrawn.

Babbage succeeded Airy to the Lucasian chair. Although he was usually right in his opinions over a seemingly unbounded range of topics, no one at government level would ever have dreamt of listening to him. Partly this was because of his social behaviour. There was nothing about him of that 'leisurely and dignified manner' Airy assumed when a colleague predicted the position of a new planet. Airy's search for it was in the event too leisurely and dignified. Neptune was eventually discovered by someone else.

In considering how governments should be advised, Plato gave due attention to conventional oracles, which he divided into two categories:

1) *sane,* based on established rules of divination and orderly assessment, and

2) *inspired,* based on the utterances of a seer or prophet.

An echo of this distinction is discernible in the practice of contemporary government science agencies, which tend to veer between foot-slogging advisory committees and the sudden inspirations of eminent individuals. We look in vain for consistent principles which might bind together different prophetic episodes, and the record of oracles in science policy, whether 'sane' or 'inspired', does not encourage generalisations. We can of course give precedence where possible to the 'sane'. But we should not overlook a sprinkling of unsatisfactory outcomes from this more cautious procedure. The British decision in 1912 to abandon all work on heavier-than-air flight, re-deploying resources into balloons, was the outcome of much patient work by Lord Esher's Sub-Committee of the Imperial Defence Committee. Documentary evidence had been sifted and expert witnesses cross-examined. Mercifully Esher had the courage to reverse himself in the following year, telling the Cabinet that he and his advisors had made a serious mistake.

Perhaps in the end the only safeguard lies in fostering, among those who may be called on to advise, the rare gift which enables a man who has gone out on a limb to say plainly and publicly 'I was wrong!'

Introductory Note to Chapters 21, 22 and 23

Readers of the previous section will know that in 1973-74 a mishap of scientific politics involved the author's laboratory and set back by many years the study of AI robotics and related fields in the United Kingdom. Officialdom has subsequently indicated eagerness to repair the damage. Bodies such as the Science Research Council may find it hard to accept some of the remedies required — notably the return from abroad and re-habilitation of some of those whose work was pilloried.

At the time, I felt amazement. Ecclesiastes, that incomparable analyst of the dark side, has warned about this feeling.

If you see in a province . . . justice and right violently taken away, do not be amazed at the matter; for the high official is watched by a higher, and there are higher ones over them . . .

A number of officials attempted at the time to palliate my wrath by explaining in just such terms some of the actions which their jobs had obliged them to take. I concluded that nothing but ignorance at the top could be the cause of the abuses, and that no good could be accomplished until the minds of public men were better informed.

The remaining articles are sampled from the many subsequently written with this idea in mind.

CHAPTER 21
Understanding the Machine

Where do we stand today in the attempt to program computers to perform difficult intellectual tasks? Have prospects improved of persuading these stubborn slaves of man to understand what they are doing?

My personal view is that the notion of 'machine understanding' is crucial not only to defining the field, but also to justifying the maintenance of such research at public expense. If progress can be demonstrated towards eliminating some of the idiot and expensive detail of step-by-step programming, then machine intelligence truly represents a novel departure in computing science, justified, putting it in its lowest terms, in the saving of programming costs. If it turns out merely to be a trendy label for recreational activities, then the boot would be on the other foot: indeed some would argue, and have argued, that the time has arrived to put the boot in!

There is a tendency among practical computer men to feel that the technology is everything and theory nothing, that theory is an excuse for woolly-minded academics to stand on the toes of the real men who are grappling with the real issues. This is as misguided as it would have been in an earlier age for the master joiners to cry that, when it comes to building a house, carpentry is all — forgetting both about the architect and about the man who is quietly inventing the steel girder.

The Science Research Council's consistent support for theoretical work on programming is among the few bright examples of government foresight on behalf of software technology — and already demonstrably paying off. I would extend the same sentiment to more speculative current work on automatic theorem-proving, closely related both to programming theory and to machine intelligence, which is pursued in a number of universities.

It is nonsense to say that such work never leads to the proof of any interesting theorem, first because this is untrue ('SAM's lemma' in lattice theory has been published in the mathematical literature — SAM is a computer program) and second because to remark that the Wright Brothers' machine never managed to fly a really interesting distance is to miss the point of their achievement.

Do we yet have any programs, even experimental prototypes, which can be said to 'understand'? The usual criteria as to whether an intelligent system (a schoolboy, for example) really understands an allotted task are (1) whether the system can answer probing questions about what it is doing, and (2) whether it can generate new hypotheses and new plans of action directed towards its goals. Interestingly enough, the above two tests indicate two areas of computer technology in which a definite building up of interest is currently apparent.

First, we have the so-called 'question-answering' systems which allow the user to interrogate large databases in something approaching English language, and to expect the system to respond flexibly, with answers which are typically deduced from what it already 'knows' — as opposed to the restrictive look-up methods of pre-intelligent software technology.

Among the large European computing concerns, Philips are known to have a special interest in this area. A prototype working system is the question-answering program and database concerning moon rocks developed by Woods of Bolt, Beranek and Newman Corp in the US. This is now used by NASA scientists on a routine basis.

A more far-reaching case of question-answering is the DENDRAL program of E.A. Feigenbaum and others at Stanford University. The user inputs to this system a mass spectrogram of unknown organic chemicals, together with ancillary information such as the relative abundance of different atoms in its composition, carbon, hydrogen, oxygen, nitrogen, etc.

The system answers in terms of the most plausible molecular graph (i.e. pattern of inter-connection among the constituent atoms) which could explain the spectral data. This program out-performs human post-doctoral chemists in certain confined areas of organic chemistry and its span of competence is being energetically widened by the Stanford workers.

At the same time, facilities for building really flexible dialogues with the system are still limited and cannot compare, for instance, with Winograd's famous program which chats in an articulate way about a simple table-top world.

The break-even point of industrial profitability for the DENDRAL project is not easy to predict, but chemists and computer scientists seem to agree that this horizon is now no more than a year or two away.

Before leaving the question-answering area of machine intelligence research, perhaps I should mention the development of 'seeing machines,' a phrase which I owe to my friend the engineer/psychologist – Richard Gregory.

In Edinburgh, as in various other laboratories of similar interests we are now able to present a computer TV system with ordinary objects (wheel, axle, piston, connection-rod; or spectacles, tobacco pipe, hammer, etc.), and to hold a limited dialogue with the system as to the identification of such objects by their visual appearance.

Very little has yet been done to 'humanise' the dialogue between seeing machines and people. But improvement of the linguistic interface is seen by most laboratories as of high priority.

In addition to their enormous theoretical interest, to psychologists and physiologists as well as to engineers, 'seeing machines' surely will have roles to play at the factory bench, whether for tasks of inspection, packaging, or automatic assembly. In this regard the recent conference on industrial robot technology organised at Nottingham by Professor Heginbotham, gave a foretaste of things to come.

Robot assistants will not earn their keep unless their supervisors can instruct them in terms more natural than FORTRAN (or even LISP or POP-2). I would go further and say that voice input through microphones to 'intelligent machine tools' will be among the immediate next moves.

The US government has already underwritten this point of view at the multi-million dollar level. An R & D programme to develop 'speech understanding' systems has been put in hand by the Advanced Research Projects Agency of the US Defence Department. The aim is to engineer over a three to five-year period, systems able to comprehend continuous spoken speech with a vocabulary of up to 1000 words. Only limited and specific domains of discourse are envisaged during the initial period, with more ambitious targets to come later.

Turning to the second of my two categories, we come to self-programming systems. By this we mean systems which are not only easy for the human user to teach by examples and other short cuts, but which are actually able to generate for themselves new programs to solve new problems. The gold rush of 'automatic programming' research shows premonitory signs in the US of growing beyond the bounds of its true strength, as occurred for a while with machine translation.

Solid achievements are nevertheless accumulating, ranging from systems able to generate simple sorting algorithms from axiomatic descriptions, to quite intricate facts of autonomous robot planning such as those under study at Stanford Research Institute. A particularly promising trend is that of 'computer-assisted programming' which does not demand that the computer does the whole job. Interesting essays along this more conservative route have been made by many people including Bob Floyd in the US, John Darlington and Rod Burstall at Edinburgh, and Brian Randell at Newcastle. The approach is nearer to orthodox computation theory than to machine intelligence, but none the worse for that.

What should we pick out of all the interesting things pursued by students of machine intelligence in different laboratories as being in some way thematic? What of the next few years?

One sure candidate is the development of integrated systems. By this I mean problem-solving software able to co-ordinate on a given task (this might be as romantic as oceanbed exploration or as humdrum as luggage-handling at airports) a wide variety of different sensory channels of information — television cameras, laser range-finders, microphones, linguistic input via teletypes, tactile sensors, etc., together with an almost equally wide variety of operational responses.

CHAPTER 22
Man's Future in the Knowledge Game

Machine intelligence is concerned with programming computers to be smarter than they are at present, particularly in areas which depend upon large bodies of knowledge. Computers still 'know' very little. They have a small repertoire of fast tricks, but they are incapable of achieving genuine problem-solving in the way that a man can. The human advantage comes from man's ability to compile and maintain in his head large bodies of systematically organised knowledge about particular problem areas.

For some very narrow tasks computing systems already out-perform the brain. We do not take great pride in our ability to do arithmetic, and so it does not bother us that computing systems are already far superior to man at arithmetic. Moving to more complex problem-solving, computational meterology is now, roughly speaking, at a level with the experts in some branches of weather forecasting. In organic chemistry, particularly in the identification of unknown chemical compounds from mass spectrogram patterns, and in some specialist areas of medical diagnosis, computers are moving into the lead.

But the socially interesting question is really a different one: 'When will it be a conceivable enterprise to develop a computing system which has an all-round capability, including a self-learning capability, which is comparable to that of a human being?' Answers to such a question are bound to be speculation: there are some who think this will have become a possibility before the end of this century. That is of course a different matter from whether there will ever be the motivation to construct an artificial all-round intelligence of that kind. My feeling is that there is no useful application which would motivate that.

The pay-off will always lie in making systems that have a great deal of expertise, superhuman expertise, in some delimited, defined domain of knowledge useful to man. There would obviously be a great pay-off in any application for which this could be done over the next few decades. To take only one example, if we could develop computing systems for really

252

intelligent and insightful economic planning, clearly this would be of extreme importance, as would the automation of scientific inference in other areas such as medicine, agriculture and engineering.

Looking a little beyond that, if in some areas of economic planning the machine systems enjoyed or acquired a reputation for greater accuracy, greater planning power and more reliable judgment than the economists could muster, then in due course this would be taken for granted, just as at present there are some branches of medicine (e.g. disease of the thyroid gland) where computing systems already make more reliable diagnoses than the expert physician. The best systems, however, are likely to remain interactive man-machine partnerships rather than stand-alone machines.

Good possibilities and bad possibilities are latent in every large-scale technological innovation. On the bad side is human dependence on increasingly computerised urban systems. At present we see the computerising of the banking systems, ground traffic control, police records, medical services, educational systems, air traffic, and so forth. Before long we will find these different computational networks interfaced to each other so that they can talk to each other and pass transactions back and forth.

We are seeing the start of this in systems which automatically debit accounts at computerised banking systems when a computer system is used to make an airline or motel reservation or a supermarket purchase. Information about citizens handled by computerised medical systems could, in other circumstances, be of interest to police computer networks.

In this last example questions of privacy are raised. But the point concerns the gradual evolution of an increasingly complex system of communication to regulate the vital functions of human cities to an extent that the complexity will ultimately escape the possibility of adequate documentation, understanding and control.

It will not be possible — it is scarcely possible now — to find any single human being who actually understands the whole of any of these large systems. Such systems might for a long time appear to be an unqualified benefit for various civic services and show gains of efficiency. The citizen in his home will enjoy connections of household to computer-controlled telephone, television and electric typewriter systems, and through these he will have access to all manner of education, entertainment and information services. But along with these gains may come an insidious debilitation of his independence.

We might slip into a rather nightmarish possibility of computer-regulated cities taking on a life of their own, with humans living in them as uncomprehending parasites. Humans might become, at best, passive pets. At worst, they might find themselves and human values — possibly even human life and health — suffering in calculations made by city control systems which could no longer be guaranteed to give priority to goals thought by humans to be important.

About possible benefits, the greatest is the expansion of cultural and education opportunities through computing systems able to do more than hand out school-room instruction in a mechanical and unappetising way. Looking to the future, intelligent systems may be evolved to act as responsive and insightful teachers, able to give individual attention to each pupil. The educational process could thus become more exciting and rewarding, motivating people to improve their knowledge and cultural background, not just in childhood and youth but throughout their lives. A hint of what may eventually become possible through audio-visual and communication technology is already visible in Britain's Open University — much envied and admired in academic circles overseas.

There is an analogy here with the Ancient Greeks, whose civilisation was made possible by slave labour. Perhaps we will see an era in which we become interested in continuing our education for life, because our society will be supported by slaves in the form of machines.

Civilisation and a high level of culture depend largely on leisure. In the past this has always meant a small leisured class. The possibility that the whole of humanity may become a leisured class is now brought into view by the further prospects of automation.

We know from past history that leisure can precipitate two quite different developments. One is on the whole good, as illustrated by the free-born Athenians who used their leisure to expand their values and knowledge. On the other hand, the ruling class of ancient Rome, another leisured class, provided an extremely different example, given up to rather crude and degrading pleasures — overeating and excitements of spectacles of mass butchery. So there seems to be no guarantee that in the creation of a leisured class, or even in the creation of a leisured species which might be the final achievement of man, we will become liberated in spirit rather than brutalised.

The social and moral issues raised by the prospect of intelligent machines are not new. Rather, the development of machine intelligence will intensify the whole range of moral issues with which we are familiar. These moral issues concern the relative rights of individuals over and above the rights of social groupings, large and small.

The point about computer power — and this is all the more true of the next phase of computer development when intelligent systems begin to make their appearance — is that there is nothing either inherently good or inherently bad about this power: what it does is to confer a vastly enhanced range of power on existing institutions.

The police will be greatly enhanced in power through having sophisticated computerised systems at their disposal. But so will the Mafia who also will have their own computerised systems. The schools and universities will be enhanced, so will the supermarkets, the factories, the law systems and the military. So also will the conservative political organisations and so will the radical forces.

All the existing institutions, which as we know have their differing goals, differing scales of moral choice and criteria, will be intensified in their capacities by this new computer revolution. So no general trend can be expected, exalting some particular moral goals and downgrading others. Rather we will see an intensification of institutional power right across the board. This could be a factor for instability in society.

As to how we should respond to research on intelligent machines, clearly I am an interested party since it is my profession. I have a natural motivation to believe that such research should be encouraged! I can attempt, however, to give a more reasoned assessment.

It is not clear, looking back, that there has ever been serious danger or damage in human history through excess of knowledge and understanding. There have been many epochs where we can see that abuses and unbearable ills were perpetuated through inadequate understanding and inadequate education. I am not aware that any new scientific revolution which has greatly increased man's knowledge and control of his environment has resulted in continuing harm, even though such revolutions engender widespread panic at the time. There are always nervous reactions aimed at discouraging sudden advances of knowledge, as in the Galilean revolution in astronomy or the Darwinian revolution in biology.

But looking back from the 20th century we can see that these panics have always been ill-founded. Nobody can now point to lasting deteriorations or dangers in human life which resulted from such great extensions of man's intellectual powers. By analogy, unless we have some special reason to think otherwise, we may expect that the same results will operate and that everything that makes man more of a master of his own destiny, more capable of understanding the forces around him and the processes of his own mind, will on balance average out to be beneficial.

On the other hand, if projections were to show that the negative effects of intelligent machines would outweigh the positive effects, and if we were really confident in the reliability of such projections, then we should even be prepared to suspend research while digesting the implications.

The generalisation about the beneficial effects of extending scientific knowledge still holds. But it *is* a generalisation, and there could be some specific area of science where, during a very critical phase of society, increase of knowledge along a particular line might be harmful.

Here is a case in point. Competent medical and biological research authorities in various parts of the world are concerned about genetic engineering: that is, deliberately re-designing the genetic make-up of certain micro-organisms. There is the possibility that as an accident, a side-effect of such research, some quite new and virulent micro-organism might multiply to an extent with which we are not able to cope. As a consequence, the Medical Research Council in Britain and relevant bodies in other nations, recently supported a six-month moratorium on research in that specific area while the matter was studied more deeply and new safeguards drawn up.

It is conceivable that machine intelligence research could at some future stage raise legitimate concerns of that character. If that ever happened then I would certainly support such a 'holding operation'.

Since the days of Galileo, man has had to revise his estimate of his place in the scheme of things several times. Let us take the Darwinian upheaval of thought as being particularly profound and also fairly recent. We now regard ourselves as not being unique living creatures but rather as being made of the same stuff and evolving through the same mechanisms as all the other living creatures in the world. From that point of view our self-picture places us in a more peripheral, less central position. At the same time, on the positive side, we see ourselves today as a much more biologically interesting organism than any 19th century man could have believed possible.

We can say the same about the prospect of machine intelligence. One day we may have to face the idea that we are not the only intelligent system on our planet, but at the same time we will understand so much more about the nature and functioning of intelligent systems, intelligent processes, that we will find our own thoughts more complex and interesting than we do now.

It is more dignified to be a 20th century citizen than to have been a Bronze Age Greek of the Homeric epoch largely because modern man has more self-confidence, is less at the mercy of forces he regards as arbitrary, incalculable and inexplicable. So in general the overall effects of machine intelligence research ought to be socially healthy. The research should be pursued, partly because of applications to various parts of the social structure and partly for knowledge for its own sake.

In conclusion, it would be wrong if an impression were created that the profession of machine intelligence scientist is in any sense peculiar or special. Machine intelligence is one among many branches of computer science, all of which are worthy of support and development. It is a particularly interesting branch because it interacts closely with other sciences, especially the psychological sciences.

The study of how existing intelligent systems, how human brains and minds function, is a point of interface with other disciplines. There have been many such interfaces in scientific history. The interfacing disciplines are often the fastest growing at any particular time.

When biology first began to link up with chemistry, biochemistry was born and this was a rapidly developing, exciting field. Machine intelligence work is currently in an exciting phase. But I would not make any special claims beyond what every scientist will say of his field, namely that he does what he does because he enjoys it and hopes it may be of benefit.

CHAPTER 23
The Only Way Out of the Economic Pit?

In the past 10 years, two developments have occurred — one in hardware, the other in programming — of such explosive potential as to change the future of computing out of all recognition.

The first everyone already knows about — the microprocessor revolution.

The second is something which a few computer scientists scattered here and there, mainly in academic laboratories, have been struggling towards for a long time. But only in the past year or so have they seen it begin to blossom into practical credibility. I refer to intelligent databases.

In 1975 only one such system of computer-embedded knowledge was operating at a level competitive with skilled human practitioners (in this case chemists). In 1976 there are at least five experimental systems deserving to be described in these terms, three of which are deployed in various sectors of clinical medicine. I would not like to predict how many such systems will be in operation next year. They will probably be numerous enough to give the machine intelligence worker a hard time keeping track of them. But thanks to the post-Lighthill policy of the Government funding agencies none will be in this country.

By intelligent database I refer to a representation in machine memory of knowledge about a complex domain of such a kind as to make possible a useful degree of problem-solving. This is, of course, no big deal unless the problem-domain happens to be 'hard'. For a domain to qualify as 'hard' it must defy both the computer man's time-worn approaches: the algorithm and the database.

To exemplify with one 'frivolous' and one serious example, it is possible to write quite a short, strictly algorithmic, program which will play chess better than Bobby Fischer. Equally, it is possible to write quite a short algorithmic program which will outperform the best chemist in the world at identifying unknown organic chemical compounds from patterns given by the mass spectrometer. Unfortunately, the running times of these programs, even if mounted on super-fast micro-micro-second computers, would be measured in multiples of the age of the Universe. So in both applications time-worn approach No. 1 is defeated.

How about doing it by lookup? Lookup tables both for chess and for mass spectroscopy are indeed theoretically possible, and are currently in use for small fragments of the respective problem-domains. But when we calculate the database sizes needed to store the complete lookup tables we find that, even after making all possible allowance for micro-miniaturisation, they would be too large to accommodate on the surface of the planet.

Conclusions: (1) chess is hard, (2) mass spectroscopy is hard; (3) to do anything interesting with either requires the building not of a lookup database but of an intelligent database. Since one of these has already been built inside every grandmaster's and every master spectroscopist's skull, we have always known that this should be possible. Now that it has actually been done for spectroscopy for the Lederberg–Feigenbaum–Buchanan group at Stanford University, we know it even more!

So let us look at a few more hard problems.

Conducting informative English-language conversations is 'hard'. So is, interpreting visual scenes, finding proofs in geometry, assessing bacteriological laboratory tests, and building assemblies by controlling 'hand-eye' devices. Those working to computerise such skills are not only motivated by the evident benefits to the computing industry and to mankind generally which are likely to accrue. They are even more strongly driven by the urge to find general laws applicable indifferently to brains and to computer programs, rather as aerodynamic laws apply indifferently to aeroplanes, birds, bats and bees and (come to that) to boats, whales and fishes.

Unfortunately the pioneers saddled this field in the early days with a name (Artificial Intelligence) so unattractive that its crucial importance to other sciences is still overlooked.

The earliest and greatest pioneer of all, A.M. Turing, had more sense. He talked about 'thinking machines'. Knowledge machines would perhaps have been even better. Apart from the jarring associations of 'artificial' with artificial legs, teeth, hair, flowers, smiles, insemination, etc., the term Artificial Intelligence implies a pretension that someone, somewhere, has already achieved something very extraordinary. Since they haven't, let us forget about this label, and think quite simply about getting knowledge and intelligence into computer databases.

The interpretation of visual scenes yields a nice example of what kinds of task need knowledge and what do not. Weather satellites circling the Earth take photographs at regular time intervals. The meteorologist would like the computer which processes the torrent of photographic data to track the 'same' cloud formations frame by frame. The appropriate computation is that of local point-to-point correlation between one photo and the next under different small relative displacements. This immense load of arithmetic may possibly be eased by use of multi-processor array computers such as ICL's experimental DAP machine.

If all turns out well we will have a neat and satisfactory solution to a pressing problem. But knowledge? Not at all! The computation can quite happily do without.

Now consider a superficially similar task. A pilotless aircraft flying at 20,000 feet is required to observe the terrain below, and to navigate using a map. The apparent similarity to the earlier task resides in the need to say that *this* structure in picture A denotes the same objects as *that* structure in picture B. In task 1, A is a photo and B is the next photo. In task 2, A is a photo and B is a map.

For the second task, no correlation method or anything remotely like it will do. This is because a TV photo of a landscape does not have any kind of point-to-point resemblance to a map. Photo and map only 'look like' each other on the basis of a great deal of (usually unconscious) knowledge-based interpretation. The human pilot knows all manner of things about what is and is not possible in a real terrain.

A well-indexed body of retrievable knowledge is necessary for tentative identifications to be confirmed, extended, abandoned, modified, and used to erect new working hypotheses. Work by H.G. Barrow, formerly of the Edinburgh FREDDY project and now at Stanford Research Institute, is addressed to this exacting knowledge-engineering task.

The most convincing and thrilling intelligent database built to date is without doubt Stanford University's Heuristic DENDRAL, initiated some 12 years ago by Nobel prize-winner Joshua Lederberg. He had in mind, and has in mind today, extra-terrestrial applications beyond the obvious earthly utility of systems smart enough to exchange ideas with experienced research chemists. Lederberg realised that when soft landings on

Mars are achieved (as has at last been accomplished by the Viking mission) a fantastic scientific opportunity apparently lies within reach — and yet in reality beyond our reach owing to the absence from Viking's cargo of a ship's scientist. Something can be done by remote control — but not too much, because of the 18-minute delay each way for transmission of signals.

Lederberg chose mass spectrometry as one of the key knowledge-based procedures to automate. It now seems certain that a Mars ship of the not too distant future will carry a miniaturised intelligent spectroscopy database to support its search for organic molecules in the red dust.

Most interesting of all is the meta-DENDRAL module of the Stanford system, whose task is to abstract from operational experience new rules for incorporation into the system itself. Since these are nothing less than new scientific hypotheses in organic chemistry, they are treated cautiously and screened by a committee of chemists before adoption. But enough rules have successfully run the gauntlet to form a substantial publication shortly to appear in the specialist chemistry literature. This must surely be the first case of an original contribution to the natural sciences from a machine!

This extraordinary feat gives a clue to ways in which the computing world may feel the impact of the intelligent database in the 1980's, namely through an intensification and broadening of the computer's role as scientist's and technologist's assistant. Other applications will of course be important, but the most technically gripping challenge, even if not immediately the most economically important, will be how to spread the computer wave from the 'front end' of the scientific process, the telescopes, microscopes, centrifuges, chromatographs, photometers, scintillometers, spark chambers and the like, back to the recognition and reasoning processes by which the chaos of data is finally consolidated into orderly discovery.

In our country's desperate plight, scientific instrumentation and knowledge has a special significance, for it may offer the single solitary rope ladder up which we might, just conceivably, climb out of the economic pit into which we have fallen. We cannot expect to compete with the Americans, the Germans or the Japanese in heavy manufacturing industry. It is going to be extremely hard to beat our rivals in electronics, more favourable for us than other manufacturing in being more brains-intensive and less capital-intensive.

But there is no obstacle other than the frivolities of our own selves to finding a niche as the world's front-runner in the design, implementation and international marketing of intelligent databases embodying the technical skills on which our inventive nation still prides herself.

Introductory Note to Chapters 24 and 25

One of the most startling presentations at the 1977 meeting in Toronto of the International Federation for Information Processing was given by Nicholas Negroponte of the Massachusetts Institute of Technology. His talk came in a session on the future of computer aided design. But the system he described adds new dimensions to current concepts of the use of computers for visualising design ideas.

Yet one person's fun may be another person's poison, as indicated in the second of these two articles.

CHAPTER 24
Pressing the Fun Button

Computer Aided Design has been called many things. To Nicholas Negroponte and his colleagues of the MIT Architecture Machine Group it is 'a self-serving exercise in fiddling with details, for the most part unrelated to what anybody would call design.'[1]

As they see it, the designer's creative role cannot be realised so long as he is confined to peering at and editing black and white drawings on conventional CRT displays. In the MIT system the user flies bodily (so his real-time sensory cues tell him) around line drawings, full colour photographs, passages of text, and anything else he cares to have the computer project from his files on to the office wall.

He can zoom right through the display to discover layer upon layer of further data-surfaces, each of which he can navigate from his instrumental chair. Using pressure-sensitive joysticks with tactile feed-back, he drives through the database much as a pilot flies an aeroplane. He can also assign his own new creations (in full colour) to this simulated environment, to which he can even listen if he wants to.

As William Donelson, one of the chief designers, points out, it would be very helpful if, for example, 'when scanning a map for a subway station one could actually hear the sounds of trains running at appropriate points on the map . . .'

All this was foretold in essence more than 15 years ago by I.J. Good with his fantasy in *The Scientist Speculates*[2] of the dentist working inside his patient's mouth. Where Negroponte breaks new ground, apart from translating fantasy into actual technology, is in seeing this multi-media enhancement of sensory experience as the doorway to new desirable relationships between people and computers and also to new relationships between people and themselves.

The basic idea of the proposed relationship is that it should present only those features which evolutionary and social experience have conditioned us to handle. How do you actually file and retrieve data? Why, says

264

Donelson, by 'remembering the letter to your secretary because it was next to the telephone, locating a stack of memos or remembering that you have to stretch out your right arm, or finding a book by noting its proximity to the large red one.'

Negroponte envisages man-machine relationships of such intimacy that he asks us to picture such a dialogue as the following, conducted on returning home (presumably to one's loved and loving personal computer) after a long and trying day:

> 'Okay, where did you hide it?'
> 'Hide what?'
> 'You know.'
> 'Where do you think?'
> 'Oh.'

Negroponte's recurrent concern is with human creativity and the deadening effect upon it of conventional computer aided design, likening the latter to talking about Cezanne to a Martian by telegram. He notes that not a single contemporary system attempts to ascertain even whether the user struggling with light-pen and light-buttons is right or left handed, and many a reader will respond sympathetically to his castigation of the 'impoverished and almost sordid nature of keyboards.'

A computing system should care enough for the user's idiosyncracies to recognise him by them. He likens such a system to a super-ideal secretary, seen as the first step towards a working environment in which the user can realise his thwarted potential to be truly creative.

I agree. The more my secretary panders to me, the more creatively I begin to behave. By the end of a good day I am inventing quite exotic things of no perceptible use. So she rations her panderings so that these creative outbursts do not in general occur on the firm's time. Probably she has in mind the anecdote of how Leonardo da Vinci devised a mechanical system 'consisting of little spoons with which different colours were to be mixed, thus creating an automatic harmony.' One of da Vinci's pupils, after trying in vain to use this system, in despair asked one of his colleagues how the master himself used the invention. The colleague replied: 'The master never uses it at all.'

Leonardo, a weekday painter, was a Sunday technologist. His aeroplanes and submarines were never built. The undertaking to divert the river Arno failed. So, one may add, did his project to develop heat-proof paints. His greatness was fulfilled only within the confining structure of his artist's training. Even mankind's greatest genius could not escape Grundy's Law that if the ratio of creativity to hard slog exceeds one part to a thousand, then the mixture is already too rich to generate tangible results.

The system which Negroponte's group is striving to call into existence — with impressive first-base success — belongs to a new species of 'Sunday systems'. Their role will be to augment those human talents whose exercise belongs to the play-days of the week. Foremost among these talents I place 'being creative.' As automation fast makes every day in increasing measure a play-day, I foresee an immense and variegated universe of mass applications. These may even include whatever ritualised gladiatorial games the nations may adopt as substitute for military conflicts made unacceptable by today's hundred-megadeath technology.

But what about the work-days? Are not 999 out of every 1000 of these best passed in studious application to the tens of thousands of tested rules which go to make each given craft?

No need to argue a point which may one day receive the verdict of the marketplace. Let the Negroponte machines of the future be equipped with a 'fun button' and a charging algorithm which discriminates between modes. I shall pay for my laboratory's 'no fun' use by selling 'fun time' to those of my colleagues in danger of taking their work too seriously.

It is only fair to add that under this rubric I shall certainly be hiring a ration of fun time for myself, if my secretary lets me.

REFERENCES

1. Negroponte, N. (1977). On idiosyncratic problems. Technical Report NN-100-1 MIT, Cambridge, Mass.
2. Good, I.J. *The Scientist Speculates*, Heinemann (no longer in print).

CHAPTER 25

Beware Scum – the Society for Cutting Up Machine-makers

Watch out for SCUM! I am not referring to the latest oil slick. The acronym SCUM was originally coined in New York by the tragic paranoid Valerie Solanas who, on June 3, 1968, shot Andy Warhol. For her, SCUM stood for Society for Cutting Up Men. With brilliant intensity her SCUM Manifesto propounded the bold idea of exterminating the male of the human species, thus freeing the world's women to develop their own destinies. Her idea was to fill the gap with machines.

My mind went back to poor Valerie's diatribe while reading Colin Hine's *The Chips are Down* published by Earth Resources Research Limited. I realised that a new kind of craziness may one day be sparked off – not man-hating but machine-hating. In seven terse and sombre pages the author warns that computer-induced mass unemployment may attain disaster proportions in Britain, and certainly will do so if a major national effort to analyse and cope is not mounted.

Hines argues that it is the microprocessor which is suddenly turning dreams into threatening reality.

What happens when people are chronically and massively deprived of jobs? My reading of history is that the very same moods of alienated frustration as those expressed in SCUM can build up quite quickly with terrorism, mass hooliganism, racial slaughter, and war not excluded in some social circumstances as final outlets. SCUM this time will stand for Society for Cutting Up Machine-makers.

Luddism never yet turned back the industrial clock. So other responses will have to be found to such facts as the following:

1) Micro-automation has already shattered the Swiss watch industry.

2) In 1970–75 National Cash Register reduced its manufacturing workforce by more than 50%.

3) In 1970–76 US Western Electric's manufacturing force dropped

TABLE 25.1

The job application form on behalf of a Unimate robot.

APPLICATION FOR EMPLOYMENT

NAME Unimate 2000 B   SOCIAL SECURITY No. None

ADDRESS Shelter Rock Lane, Danbury, Connecticut 06810

AGE 300 hours (by software extension — 15,000,000 hours)

SEX None   HEIGHT 5ft.   WEIGHT 2,800 lbs.

LIFE EXPECTANCY 40,000 working hours (20 man-shift years)

DEPENDANTS Human employers of Unimation Inc.

NOTIFY IN EMERGENCY Service Manager, Unimation Inc. (203/744-1800)

PHYSICAL LIMITATIONS Deaf, dumb, blind, no tactile sense, one armed, immobile

SPECIAL QUALIFICATION Strong (100 lb. load), untiring 24 hours per day, learn fast, never forget except on command, no wage increase demands, accurate to 0.05″ throughout sphere of influence, equable despite abuse

HISTORY OF ACCIDENTS OR SERIOUS ILLNESS Suffered from Parkinson's Disease (since corrected), lost hand (since replaced), lost memory (restored by cassette), hemorrhaged (sutured and fluid replaced)

POSITION DESIRED Die cast machine operator

OTHER POSITIONS FOR WHICH QUALIFIED Forging press, plastic molding, spot welding, arc welding, palletising, machine loading, conveyor transfer, paint spraying, investment casting, heat treatment, etc.

SALARY REQUIRED $4.00/hour

RELATIVES IN THIS PLANT Five 2000A Unimates in forging department

LANGUAGES Record-playback, assembly, Fortran

EDUCATION On the job training to journeyman skill level for all jobs listed above

REFERENCES General Motors, Ford, Caterpillar, Babcock Wilcox, Xerox and 65 other major manufacturers from 39,200 to 19,000. A 75% cutback is additionally expected in fault-finding, maintenance, repair and installation.

4) Similar trends are now apparent in the service industries through micro-automation in supermarkets, in the garment trade, in offices, in design shops and elsewhere.

5) Overall, unemployment in Britain is projected at the level of 5 million by 1990.

Hines also takes note of the growing impact of the reprogrammable robot. There have been recent references to the mushrooming US robot manufacturer Unimation. The job application form shown has been filled in on behalf of a Unimate robot. The corporation puts it out as a promotion gag. But the joke has disturbing overtones. Valerie Solanas' fantasy of replacing male persons by machines acquires a hint of feasibility – but females are equally vulnerable. A recent Times articles described a word processor as, 'The four thousand pound typist-substitute that will soon pay for itself.'

Where do we stand as computer people? We started it, didn't we?

The situation is more acute because there are not one but two problems building up: (1) job-destruction, (2) the need to fight overseas competition both by modernising our own plant and by developing automation systems for sale overseas.

Yet (2) seems to aggravate (1). Well may Hines say 'There will be no easy, wand-waving solutions'!

All that a computer man can do is to see if his own professional skill can be brought to bear on some sector of the problem. For example, contributing to 'the human interface', so as to make obscure brute-force computations more comprehensible to the machine's supervisors, is clearly one of the more constructive directions. Allied to that, the whole area of computer aided education and training acquires a new, and especially beneficial, significance.

One must sympathise with union pressures for further shortening the working week. None the less, I prefer the 'four plus one' day to the 'four-day' week. The 'one' stands for all-day educational release at employers' expense. Such a scheme would work for the education industry, which itself can absorb with benefit indefinite amounts of automation to augment, not displace, instructors.

If we computer men do not concern ourselves about the good and evil we sow, then a day may dawn when we do indeed have to watch out for SCUM.

REFERENCE

Solanas, Valerie (1968). *SCUM Manifesto*, New York: Olympia Press.

Introductory Note to Chapters 26 and 27

Machines unable to explain themselves are part of life's wearisome side. It would be nice, of course, if our car could tell us the cause of the latest ominous rattle or clink. It would be nice if our hand-held calculator could defend or retract its apparent belief that $\sqrt{-x} = -\sqrt{x}$. But when the machine system grows beyond a certain complexity, and has responsibility (let us say) for the automatic operation of a steel-strip rolling mill, for the monitoring and drug-management of patients in a cancer clinic, or for the control of the air traffic over a large city, then the issue of mental *rapport* between intelligent machine and its intelligent user is transformed from one of convenience to one of life and death.

This article explains the nature of 'conceptual interfaces,' the construction of which (so I believe) is among the most urgent tasks before us. In *Social Aspects of Artificial Intelligence*, the last of the collection, this theme is broadened into a survey of the unprecedented hazards and opportunities which are expected increasingly to dominate the closing years of this century.

CHAPTER 26
The Human Interface

The computer revolution is about to pass a critical threshold beyond which lies danger. On the safe side of this threshold are systems which cannot explain their own actions, reactions or inactions — yet explanations *can* be found if the user is determined enough to invoke a repair shop, or a manufacturer's design manual, or some other source of diagnostic insight. Computing systems operating on the *far* side of the threshold, on tasks of great complexity, are in an entirely different case. With one saving proviso, to which I shall come shortly, nothing in the world can then explain these machines' operations. Users may want to know whether the system knows what it is doing because they are frustrated (e.g. it is a scheduling system for finding trans-Continental motoring routes), or anxious (e.g. it is a previously untried steel-mill automation system), or in danger (e.g. it is an air-traffic control system). They will need something more re-assuring than (a) a print-out of the program together with the last few hundred thousand operations which it has just performed, or alternatively (b) the relevant entries in some vast stored database of pre-calculated solutions.

The saving proviso is the possibility, by building conceptual interfaces for such systems of bringing the problem-solving mentalities of man and machine into closer *rapport*. Unfortunately there is no way of doing this, short of implementing in the machine a conceptual model of the problem-domain essentially similar to that possessed by the human. Explanations are of no use unless they are expressed in concepts familiar to the receiver of the explanation.

Is it possible to program a computer *in concepts* instead of in the full detail of individual operations? The essence of John McCarthy's 'Advice Taker' proposal, advanced over 20 years ago, was that this should be possible, and also that if computers are ever going to be able to work with us in a really intelligent way we must find out how to program them at the conceptual level.

272

WHITE TO PLAY
No match with rule CR. No match with Rule 1.
Position matches rule R2 (our K on edge, our K and N not separated).
Rule R2 says try advice nos. 1, 5, 6, 7, 8, 11 and 12.
Advice 1 "KILLROOK" found unsatisfiable.
Advice 5 "HOLDEDG1" found unsatisfiable.
Advice 6 "HOLDEDG2" found unsatisfiable.
Advice 7 "HOLDEDG3" (namely avoid mate, avoid loss or danger to knight, keep K and N together) *is* satisfiable by moving **K-R8.**
BLACK REPLIES **R-QR2**

Position after Black's move R-N2 ch.

WHITE TO PLAY

Position matches rule CR (special pattern with K in corner).
Rule CR says try advice no. 9.
Advice 9 "CORNCASE" (namely make the distance between our K and N equal to 4) is satisfiable by moving **N-N4.**
BLACK REPLIES **K-B2**

Position after Black's move R-QR2

WHITE TO PLAY

No match with rule CR. No match with rule R1. Position matches rule R2 (our K on edge, or K and N not separated).
Rule R2 says try advice nos. 1, 5, 6, 7, 8, 11 and 12.
Advice 1 and 5 found unsatisfiable.
Advice 6 "HOLDEDG2" (namely avoid mate, avoid loss or danger to knight, maintain or increase distance of our K to corner, our K to be not more than 3 squares from N, our N not to be adjacent to their K) *is* satisfiable by moving **K-R7.**

Position after Black's move K-B2

FIGURE 26.1. An episode during play of the king-knight-king-rook end-game by an 'expert system' against opposition provided by a US National Master who plays the rooks's side (Black). White's knight move is counter intuitive but necessary. Commentary is an edited version of a machine-generated explanation of its own play.

In recent years real progress has been achieved and 'expert systems' for fields as diverse as geology, chemistry, medicine, agricultural botany and chess are proliferating in laboratories studying machine intelligence, mainly in America. A key feature has been the development of new computer languages for knowledge representation to which I have given the generic name 'advice languages'. Figure 26.1 illustrates the performance of a computer which has been taught, using our Edinburgh advice language, how to perform a complicated task in chess (defending with King and Knight against King and Rook). Since its internal representation of its task uses familiar chess concepts it is able to give a stage-by-stage account of what it is trying to do.

It is possible at any moment to interrupt such a program and give it fresh advice. Alternatively, having gradually polished up its internal representation through accumulation of trial and error, one can translate it back into human-readable form. This produces a more precise and complete codification of knowledge for the given task area than existed before. Table 28.1 shows a list of application areas from which computer-assisted 'knowledge refining' of this kind has been reported.

After a somewhat lengthy period of incubation there are signs of new initiatives in a number of British institutions with responsibility for computer science and technology. I am one of many in the field who believe that the need for safe, because explainable, machine systems is already looming large and that the UK should aim to be in the forefront of the requisite new developments. Suggestions have even been made for a 'national project' focussed on a target carefully selected so as to force the pace in precisely the needed areas.

Although there are reservations to be expressed, history suggests that very occasionally a critical selection of target can by good fortune be achieved. During the war an artificial target generated from cryptographic needs led to a unique concentration of high-speed electronic computing know-how in British hands, augmented by related know-how acquired in war-time radar, from which a short-lived technological hegemony was established in the early post-war years. The lead was eventually lost, but from political and financial causes, not technical.

An analogous case may be being enacted at the moment. Although NASA was not moved by the desire to give the USA pre-eminence in exploratory land vehicles, this will be the long-term consequence of their specifying and placing with the Jet Propulsion Laboratory at Pasadena the 'Mars Rover' contract. Since the surface of the earth is not covered with highway, much of it being at least as adverse as the boulder-strewn Martian terrain, we can be sure that the high-technology corporations who are now taking JPL sub-contracts can count on dependable terrestrial applications even if the planet Mars were to fall out of the sky.

It can be said that these examples were in a sense accidental. The artificial target was not deliberately constructed for applying a forcing function to new technique. Two answers: (1) there has to be a first time; (2) it had

TABLE 26.1

Expert systems as sources of improved codifications of human knowledge

Domain	Previous codification	Refining instrument	Desired end-product
Chess: spotting mates 'at a glance'	No non-trivial classifications published	PL1 tournament program MASTER	Reference text of mating patterns
Chess: how to mate with king and rook against king	Chess primers by Capablanca, Fine, etc.	AL1 'Advice Taker' program	Six sufficient rules, formally proved correct
Chess: how to defend with king and knight against king and rook	Chess primers by Fine and Keres	AL1 'Advice Taker' program	Micro-manual of pattern-based rules
Chemical synthesis planning	Text books on synthesis	SECS program with data-base of chemical 'trans-forms'	Improved source of synthesis-relevant knowledge for chemists
Planning robotic assembly sequences	Toy car assembly scheme for Edinburgh versatile assembly	PROLOG predicate calculus-based program	Improved assembly sequence
Plant pathology	Pathologist's diagnostic Classification of soybean diseases	AQVAL program for inductive inference	Improved set of classificatory rules
Mass spectral information on mono- and poly-ketoandrostanes	No satisfactory pre-existing explanation of spectroscopic behaviour	Meta-DENDRAL module of DENDRAL program	Substructures defining main cleavages, yielding predictive theory for new ketoandrostanes

Preliminary indications can be discerned in the above cases that a codification of a given skill by a human expert can be 'refined' into more exact, complete and reliable form by use of appropriate expert systems.

better be soon. The acceleration of technical possibility is such as to compound our vulnerability should some competitive nation begin to frame such targets.

Can an appropriate target, then, be framed within the context of the new initiatives?

The end-system should point towards 'knowledge-intensive' rather than 'capital-intensive' technologies. These should be relevant to social problems of automation and of the post-industrial society.

It may be asked, what is wrong with the most obvious and immediate idea to suggest itself, namely a prototype fully automatic factory? Other advanced nations are seriously considering the launching of such projects. I believe that to scramble blindly for bigger, better and more total automation could be to join a Gadarene rush. The greatest social urgency attaches not to extending automatic processes but to *humanising* them. We have at the same time to build on the notion that the primary manufactures of the 21st century will not be material but informational goods.

An alternative target for Government-aided research, and much more socially relevant, would be a *knowledge-engineering workshop*. It should be tooled, staffed and organised so as to function as a model facility dedicated to the design, development, testing, documentation and packaging of a new kind of intellectual product — namely advice languages, and advice-machines built with their aid. These are 'intellectual products' in the same sense that academic text-books and industrial maintenance manuals and user guides can be so described. There is an additional reason, namely, that *active intellectuality* forms part of the product's functional specification!

Specimen demonstration versions might be the following:

1) a chess-player's assistant which a 'Class C player' can use to become after a few months' practice the equal of a grandmaster when using it in man-machine mode.

2) a chemist's assistant which a trainee chemist can use to become after a few months' practice as valuable to a chemist company when he operates in man-machine mode as a senior chemist.

3) a clinician's assistant which a junior houseman can use to become after a few months' practice as valuable to the hospital when he operates in man-machine mode as a senior consultant.

Other expert domains suggest themselves, such as computer-aided design, economic forecasting, geological prospecting, disaster advising, and many others.

An important effect of the development of each advice-machine representation will be a new and much improved theory of the corresponding domain of expertise. Uses of advice machines will be:

a) to act *autonomously* as question-answerers and problem-solvers;

b) to be used *interactively* for question-answering and problem-solving;

c) to be put to conscious *tutorial* uses (unconscious tutorial effects are inseparable from (b) above).

Item (b) is directly relevant to a new need which is beginning to surface, and which threatens one day to become widespread. Prodigious resources of brute force computation can now be unleashed on complex problems of prediction and control. Result: the human partners in the decision-making process (shop-floor supervisors in automation systems, control-tower personnel in air-traffic systems, doctors in automated patient monitoring and treatment) can lose their nerve through finding the machine operations insufficiently comprehensible. Lacking a *conceptual interface* to help them see 'what the system is trying to do' at each stage, responsible staff have been known to retreat into dangerous and costly passivity. Study is needed of model systems addressed to this danger. We need to explore the use of advice machines as interpreters for conceptually opaque problem-solving systems.

Bringing the concept of a knowledge-engineering workshop to viability cannot be done at all without a working unity between certain kinds of academic specialist and a spectrum of more practical skills. Unusually good opportunities for such unity exist today.

CHAPTER 27

The Social Aspects of Artificial Intelligence

Introduction

The latest advances of 'computer-on-a-chip' technology have caught the public eye only recently, although the technology has been in use for some time. A time-sequence of effects on society may be listed thus:

1) Industrial robotics;

2) Electronic toys (including educational);

3) Office automation;

4) Home hobby computing;

5) 'Expert systems'.

Progress in areas 1-3 is already well under way, and an increasing number of hobbyists in the UK are now using computers at home.

Discussion of 'Artificial Intelligence' (AI) in this article will mainly be concerned with elaborations of theme 5. An expert system is a computing system which embodies organized knowledge concerning some specific area of human expertise (medical diagnosis, chemical identification, economic geology, structural analysis, number theory, chess, etc.) *sufficient to be able to do duty as a skilful and cost-effective consultant.* Seemingly non-economic fields, such as number theory or chess, do in fact generate tuition and consultation fees. So even here the 'cost-effective' part of the above definition is applicable.

The effect on AI, defined in this narrow way, of the declining costs of computing hardware has been rather suddenly to bring the economics of expert systems above the threshold of cost-effectiveness. The PROSPECTOR program[1] developed at Stanford Research Institute for advising mining companies when and where to drill for ore already costs only ten dollars worth of machine resources for each consultation dialogue. Such tasks cannot be done without machine embodiments of processes which we are accustomed to describe as 'intelligent' when we see these being executed by ourselves or by others; for example, by human experts who may be hired for their advisory services. Questions put to such an expert, or to the 'expert' machine, are not in general susceptible of solution either by lookup of pre-stored answers or by simple calculation. Answers require reasoning, search, pattern-matching, acquisition of new concepts, judgements of likelihood and revision of judgement in the light of new data. In short, they require intelligence.

If present successes develop along the rising curve followed by other new technologies, then it is evident that serious social consequences may follow from endowing machines with these facilities.

There is a need to develop some documented approach to AI, to its likely lines of development and future impact. We need to avoid the mistake of the physicists in the 1930s who were *laissez-faire* about the implications of what they were doing. Ernest Rutherford stated as late as 1936 that he saw no possibility of either commercial or military exploitation of atomic energy. In very recent times another branch of science has arisen, genetic engineering, with considerable consequences for society, not fully foreseeable. The geneticists have done better than the physicists, who left the politicians facing awesome problems under circumstances of rush and secrecy without any adequate file of technical information. Some years ago geneticists in all countries called a halt to recombinant DNA experimentation within certain categories, and had a moratorium for six months, in order to give time for risk-assessment to catch up with laboratory technique. That moratorium has since been lifted, but in the interim detailed ground-rules have been drawn up and are now operational.

In 1972 I convened an international panel of AI scientists to consider whether a similar situation might one day arise in our own field. We spent three days as guests of the Rockefeller Foundation at the Villa Serbelloni in northern Italy. A 'Serbelloni file' of a sort now exists.

One thing we had to consider was whether our study was premature. We looked at three possible answers to this. One was that it is far too early in the present state of knowledge to worry about this issue. The extreme opposite view is that social consequences are already knocking on the door. We settled as of then for an intermediate assessment. The following is taken from the letter that we sent to the Rockefeller Foundation.

It is clear that a wide range of increasingly important social consequences can now be seen to be following developments in the field of computer science and automation. Machine intelligence, when it is developed to any significant degree, may add a new dimension to these capabilities, and therefore may have the effect of increasing the scale of the above social consequences and of raising quite new issues.

What is new in the eight years since then? The new factor, according to the argument presented in this chapter, is the appearance on the scene of a growing number of 'expert systems'.

New Paths in Information-processing

Some systems, such as de Dombal's program for diagnosing acute abdominal pain,[2] have a simplistic and uniform representation of the processes of knowledge-based inference, and this limits the depth and quality of the 'self-explanation' facilities which can be provided, and also the degree to which the system itself can actively and intelligently seek from the user additional facts to assist its deliberations. More advanced systems, such as the MYCIN antibiotic therapy counselling program, are organized in a more structured fashion, reflecting in fair degree the manner in which the human specialist arranges, and makes inferences from, his knowledge of the subject. Common to all representations is a departure from the accepted norms of standard programming practice. As discussed early in Chapter 16, work with expert systems has posed a new challenge to computer science — namely the design and implementation of linguistic vehicles suitable for the domain-specialist. Such languages are called 'knowledge-representation' or 'advice' languages.

The critical factor affecting the speed at which the new technology can take off is the degree to which expert systems can be made capable of adding to, or refining, their own rules in the light of examples and counter-examples, either presented by the user in tutorial mode or, in simple cases, directly from the data.

Machine Learning

In an expert system such 'learning' takes the form of the program's augmentation, or modification by simplification or correction, of its own stored rules. An early laboratory demonstration was a program[3] which by trial and error acquired the ability to balance a pole on a motor-driven cart running to and fro on a track of fixed length. This task is near the margin of tractability for models based on conventional adaptive control theory and these yield systems which are computationally relatively costly at run time. The program was called BOXES. It was able to operate on a basis of 225 rules, and provided a simplified illustration of what has some-times been termed the 'Committee of Experts' principle. A Chairman (central control routine) after inspecting each input state invokes from a library of problem-solving rules whichever is known to contain specific 'expertise' concerning situations of that particular type. A 'situation' in the pole-and-cart task might take some such form as 'Cart near left-hand edge of track, cart moving to the right, pole moderately inclined to the right, pole swinging to the left'. Once invoked, the rule prescribes a burst of power from the motor either to the left or to the right, and a new input state is generated. The collective expertise of all the rules put together determines the quality of system performance. At the start of a learning series the 'left' and 'right' decisions were set at random over the set of 225. With accumulating experience the system learned to perform as an expert pole-balancer.

BOXES was essentially statistical and lacked hierarchical structure for expressing logical and causal relations. The same can be said of the de Dombal program for diagnosing acute abdominal pain. Both programs, however, were able to out-perform the general run of human practitioners. The question arises whether an improved learning model would lead to yet higher performance, and also permit machine acquisition of skills too com-plex for such simple statistical approaches.

Visual recognition by the Edinburgh robot FREDDY[4] introduced a somewhat more sophisticated representation. Users were able to teach the system the look of a given kind of object by showing examples. The properties-and-relations graphs employed to express abstracted descrip-tions were similar to the 'semantic nets' now in widespread use. More general still are *inductive inference* systems which represent and up-date their stored knowledge in the form of predicate logic sentences. A recent example of computer induction in a system for diagnosing soy-bean diseases is illustrated in Figure 27.1.

AQ11 in PL1;   program occupied about a million bits of store.

SOY-BEAN DATA:	19	common diseases;
	35	basic descriptors, size of value-sets ranged from 2 to 7;
	307	cases presented as initial training set in the form of descriptor-lists paired with confirmed diagnoses.
Test set:	376	new cases.

Machine trials using rules of three different origins.

{ 83% accuracy using Jacobsen's rules,

93% accuracy using interactively improved rules

99% accuracy using rules inductively generated by the AQ11 program.

FIGURE 27.1.   Experiments with computer induction by Chilausky, Jacobsen and Michalski (*Proc. Sixth Annual International Symposium on Multi-valued Logic*, Utah, 1976).

Emycin: A Framework for Computer Consultation

A related aspect of computing in which Artificial Intelligence methods play a central role is the incorporation of 'general-purpose' capabilities so that learning and problem-solving power can be switched from one domain to another. The EMYCIN domain-independent system for representing knowledge is a case in point. It was developed from the domain-specific program MYCIN for clinical consultations in infectious diseases, originated by a medically qualified computer scientist, E.H. Shortliffe, working at the Stanford Medical Centre in interaction with Stanford's Heuristic Programming Project.[5] As input it takes data from bacteriological tests on blood and urine samples, together with other information about the patient volunteered by the user or extracted from him in dialogue. The English-language interface sustains conversation at a semi-literate but acceptable level.

The performance of the program for a given group of diseases seems to be chiefly limited by the quantity of knowledge which it has been thought worth taking the time to impart in tutorial mode. In the case of meningitis sufficient tuition of this kind occurred to consider subjecting the program to a full-scale clinical trial of the quality of its prescribing behaviour.[6] Ten diagnostically challenging cases were selected under the constraint that there were to be no more than three cases of viral meningitis, and that there was to be at least one case from each of four categories: tuberculous, fungal, viral and bacterial. Detailed clinical summaries were presented to ten diagnosing-and-prescribing agents, namely the program together with nine medical personnel at various levels of experience. The prescriptions thus obtained were then given to a panel to evaluate consisting of meningitis experts. The identities of the ten prescribers were concealed. The highest mean rating was scored by the program.

By stripping out all the medical knowledge from MYCIN, an empty framework called EMYCIN ('Empty' MYCIN) was obtained. Its adaptability as a general-purpose knowledge-handling tool was tested with a problem encountered by a company marketing a large library of structural analysis programs for engineers. When used expertly, the library had proved its worth, but it was found that users required about a year of fairly continuous use before becoming fully expert. The proposition, then, was that EMYCIN could be used to build a system which could give expert advice to a library user as he worked his way though a problem.

The resulting systems, SACON, was built using EMYCIN in six man-months, and is undergoing field tests.[7] This experience may fore-shadow an important use for expert systems, namely in helping to establish user-friendly conceptual interfaces to mediate human access to increasingly complex and opaque computing libraries and systems.

Another unexpected direction for knowledge engineering was indicated by MYCIN. At a time when the program's clinical expertise was too fragmentary to make it of interest to practising doctors, it was observed that it was in frequent use by medical students, including as a source of instructional rules. It seemed that, fragmentary though MYCIN's knowledge was, its orderly arrangement and field tested (i.e. 'conceptually de-bugged') status made it a more useful source for some areas of knowledge than the available medical textbooks.

The implication of this anecdotal finding has subsequently been confirmed in a variety of domains. The possibility can be inferred of a new computer-assisted craft in which expert systems are used as 'refineries' of human book-knowledge. These could bear the same relation to pre-existing codifications as oil refineries bear to crude oil. In the previous chapter some illustrative instances of this 'knowledge refining' phenomenon were tabulated.

Calculation Versus Pattern-Knowledge in Problem Solving

To achieve software representations of highly trained intellectual performance some acquaintance is necessary with how the brain goes about it — not in neurological but in information-processing terms. This is however made more difficult by an almost universal delusion that when acting as a vehicle for complex problem-solving the brain conducts itself in the style attributed to computers. Actually by computing standards of sheer memory and calculational capacity the brain is extremely feeble (see Figure 27.2). Hence the virtuoso performance, whether of 'calculating prodigies' or chess-masters or any other giants of intellectual life, must be built in another way — a way which somehow compensates for man's tiny working memory and lumbering processor.

From detailed observations and analyses performed by psychologists a common picture has emerged. This picture bears striking similarities to the pattern-based rules and strategies of search and control which the knowledge engineers, working along their different path, are beginning to incorporate into expert systems. Rough estimates[8] of the number of brain-stored patterns underlying chess-master skill fall in the range 10^4 - 10^5, in harmony with estimates made in other intellectual domains.

These findings have a relation to phenomena of visual perception, already known to be largely driven by pre-stored patterns. Recent work with children's drawings[9] has reinforced these ideas. In the interpretation of the flood of optical information which pours into the human eye, or the computing system's TV camera, the critical filtering role is played by a massive battery of pre-stored *knowledge* about how the world looks, some at least of this knowledge being coded in symbolic rather than directly pictorial form.

For purposes of defining and measuring knowledge we find that classical information theory has left a gap in connection with the intuitively perceived differences between 'knowledge' and 'information'. By straightforward application of existing information theory the gap can be bridged and numerical explications can be obtained of 'problem-difficulty', 'knowledge-content' of a program, and cost-benefit of problem-solving representations.[10] Moreover it becomes possible to identify what features make a problem a 'standard' problem or an 'AI-type' problem relative to the properties of the intended solving device. A key notion which emerges from such analysis is that of the 'human window'. If machine representations are more 'bitty' and data-base-orientated than corresponding human representations of the same material, then they become opaque to the

1	Rate of information transmission along any input or output channel	30 bits per second
2	Maximum amount of information explicitly storable by the age of 50	10^{10} bits
3	Number of mental discriminations per second during intellectual work	18
4	Number of addresses which can be held in short-term memory	7
5	Time to access an addressable 'chunk' in long-term memory	2 seconds
6	Rate of transfer from long-term to short-term memory of successive elements of one 'chunk'	3 elements per second

FIGURE 27.2. Some information parameters of the human brain. Estimation errors can be taken to be around 30 per cent. (Main sources: Miller (1956) *Psychology Review*, **63**, pp. 81–97. Stroud (1966) *Ann. N.Y. Academy*, and sources cited by Chase and Simon (1974) *Cognitive Psychology*, **4**, pp. 55–81.)

human user. Equally, a machine representation can fall on the other side of the window, being too compact and processor-oriented. In applications where man-machine rapport is critical, either type of representation, however operationally efficient, can be unsatisfactory and even dangerous through failure of the user to fathom the operations which he is supposed to be monitoring. Artificial Intelligence systems by definition are those which fall *within* the 'human window'.

Plausible Inference

There is a need to extend analysis to systems which reason (as do humans) in probabilistic as well as in deductive style and which store and manipulate systems of *belief*.

A program of this type, PROSPECTOR, developed by Hart, Duda and Einaudi uses a base of stored rules and the facts elicited during consultation to advise economic geologists what is the likely pay-off from drilling for a stated mineral. Currently PROSPECTOR has only five models:

1) Kuroko-type massive sulphide;

2) Mississippi-Valley-type lead/zinc;

3) near-continental porphyry copper;

4) Komatiitic nickel sulphide;

5) sandstone uranium.

In these narrow domains the program performs comparably with human experts. Cost of a typical consultation session is about \$10. As with MYCIN, increasing the program's breadth of expertise seems only to require investment of time spent by geologists in tutorial mode. PROS-PECTOR also resembles MYCIN in that its domain is thoroughly 'messy' and hence requires a methodology for drawing approximate inferences with stated levels of confidence from incomplete and inexact data. Rather than use an *ad hoc* system for combining and weighting inferences, PROS-PECTOR bases itself on Bayes' system of inverse probability.[11/12] The de Dombal system discussed earlier does the same, though its lacks the rich structure of PROSPECTOR's 'inference net'. Incorporation of processes of plausible inference into expert systems have both improved the quality of reasoning and also enriched the facilities for the system to explain and quantitate its beliefs in response to questioning by the user.

Forecasting

There has not been much systematic forecasting from within the AI pro-fession. The best-known exercise was the 'Delphi' forecast published in 1973 by two scientists in Lockheed working with two in Stanford Re-search Institute.[13] Use of the Delphi method is fairly widespread in fore-casting the rate at which technologies needed to develop new products are likely to come along. Where this method is applicable, as in some branches of industrial technology, the track record is good.

In the AI case, the main results are shown in Figure 27.3. Nothing has happened or failed to happen in the intervening seven years which seems at all inconsistent with this set of forecasts. There is one item, a rather curious one, on which the authors lose — not in estimating the date of the given technology but in estimating the likely commercial importance. Universal Gameplayer, at the bottom, is one of the low-importance entries. But it now looks as though it is not working out that way. Microprocessors are beginning to be sold to the householder to clip on to his television set.

Products	Median Proto-type Date	Median Com-mercial Date	Desir-ability (*)
High potential significance			
P5 —Automatic identification system	1976	1980	+
P8 —Automatic diagnostician	1977	1982	+ +
P13—Industrial robot	1977	1980	+ +
P1 —Automated inquiry system	1978	1985	+
P9 —Personal biological model	1980	1985	+ +
P11—Computer-controlled artificial organs	1980	1990	+
P18—Robot tutor	1983	1988	+
P16—Insightful economic model	1984	1990	+ +
P2 —Automated intelligence system	1985	1991	0
P20—General factotum	2000	2010	+
Medium potential significance			
P14—Voice response order-taker	1978	1983	+
P3 —Talking typewriter	1985	1992	+
P15—Insightful weather analysis system	1980	1985	+ +
P6 —Mobile robot	1985	1995	0
P4 —Automatic language translator	1987	1995	+
P12—Computer arbiter	1988	1995	+
P10—Computer psychiatrist	1990	2000	0
P17—Robot chauffeur	1992	2000	+
P21—Creation and valuation system	1994	2003	+
Low potential significance			
P19—Universal game player	1980	1985	+
P7 —Animal/machine symbiont	2000	2010	−

(*) + + Very favourable
   + Favourable
   0 Balanced
   − Detrimental

FIGURE 27.3.   Summary of Delphi survey on anticipated time-scale for development of various forms of intelligent systems.

What is he paying for? High on the list is the facility to play all sorts of games. So the forecasters were right in including this as spin-off from AI research, but quite wrong in thinking that it would be commercially unimportant. It may, for all we know, become the drug of future generations in something like the way that chariot-racing and similar activities dominated the Roman, and later the Byzantine, Empire, for century after century of settled life.

Benefits to Science

Where will artificial intelligence take us from the social point of view? Let us start with benefits. First, an important predictable benefit is to the scientific community. There are two aspects of this. First, scientists, as a matter of empirically established fact, do not read (not that they cannot, but they do not). A survey was commissioned in 1956 by the UK Department of Scientific Industrial Research[14] to find out what are the channels of technical information from which scientists draw. It turns out that reading technical papers and books (not skimming, not going to the library and running through the titles in the journals, but sitting down and reading connected prose) is something to which the scientist devotes little time. On the other hand he spends much time on alternative methods of gathering information. He thumbs rides on other people's mental work. He calls up a colleague or walks down the corridor. He will go to enormous lengths to get the information out of a 'question-answering machine made of meat' to paraphrase Minsky, rather than be driven to read it up for himself. What is more, he is right. The scientist, like everyone else, is busy. He insists on his information being as much as possible *pre-digested*. The difficulty and inconvenience of retrieval from the raw state is such that scientists go out of their way to avoid it. In the long run the only technology which can break that barrier is artificial intelligence, the artificial librarian's assistant. The truly intelligent terminal for interrogating library catalogues could make the business insightful, as today the librarian does.

Suppose that we extrapolate and say that there are about twenty expert systems in operation at present, and that there are certain to be a few score in a year or two's time. A natural notion comes to the surface. Should we set up integrated reactive libraries, integrated knowledge banks? Even the question of maintaining the new experimental systems on some centralized, rationalized basis might have its attractions. For the first time in science we have a branch of scientific activity whose major goals and products also constitute new instrumentation for the subject. An information scientist who wants to work on large knowledge systems will want to go to a place or places where there *are* large knowledge systems which are in an active state of maintenance, metabolism and extension. He will be gravitating to such centres not because of interest in the expert subject matters of those large systems, but as an information scientist. In much the same way the nuclear physicist gravitates to CERN or Brookhaven where there is sophisticated, well-maintained instrumentation.

World Brain

This idea was anticipated, like much else, by H.G. Wells. In 1938 he published a book of lectures called *World Brain*.[15] A striking aspect of the book is there there is nothing in the information technology of the 1930s that gave any kind of credible basis for it. Yet he had the concept that there should be set up a 'world brain' consisting of a task-force of trained information scientists with all the information-handling marvels of their era and that they act as a kind of planning laboratory for the world. They would be the people who knew what was going on, digesting and sifting and interrelating man's rapidly expanding horizons of knowledge. H.G. Wells' world brain is now technically feasible. But is it desirable? Is it likely to happen?

There is a reasonable argument that it is not likely in the centralized form he envisaged. Consider the US telephone system. It is highly *decentralized* in its management. Yet the entire system uses Bell Lab's technology. In order to ensure an even spread of knowledge-engineering techniques, it is not essential to have either a bureaucratically or a geographically centralized structure for interconnecting different knowledge systems. That is the moral to be drawn from the example of Bell Telephone Systems. But the moral also includes the desirability for AI of a centralized equivalent to Bell Laboratories, even though the technology fed by it is most unlikely itself to be centralized.

General Benefits

So much for the scientists. What about other benefits? This is part of a broader issue concerning advanced computer technology. I was delighted to come across Newell's lunch-time speech at the inaugural party for the Whittaker Professorship in Carnegie-Mellon University.[16] His speech was called 'Fairytales'. He uses fairytales as a fanciful and charming paradigm for technology. He says:

I see it differently. I see the computer as the enchanted technology. Better, it is the technology of enchantment. I mean that quite literally, so I had best explain. There are two essential ingredients in computer technology. First, it is the technology of how to apply knowledge to action to achieve goals. It provides the capability for intelligent behaviour. That is why we process data with computers – to get answers to solve our problems. That is what algorithms and programs are all about – frozen action, to be thawed when needed. The second ingredient is the miniaturization of the physical systems that have this ability for intelligent action. Thus, computer technology differs from all other technologies precisely in providing the capability for an enchanted world. The little boxes that make out your income tax for you – the brakes that know how to stop on wet pavements – for instruments that can converse with their users – for bridges that watch out for the safety of those who cross them; for street lights that care about those who stand under them, who know the way so no-one need get lost. In short, computer technology offers the possibility of incorporating intelligent behaviour in all the nooks and crannies of our world. With it we can build an enchanted land.

He goes on also to consider some of the dangers, but believes that they can be overcome.

We must distinguish (and here I include intelligent systems of the future), between

1) *amplification of existing institutional power and activity;*

2) *the appearance of fundamentally new benefits and dangers.*

Dangers of Amplication

Let us first consider just amplification of good and bad things which exist already. Of the dangers, the most obvious is the military danger – the unmanned systems such as the Cruise missile, described in some detail in a *Scientific American* article.[17] The device has a stored model of the terrain intervening between source and destination, and by continually bouncing signals off the ground below it, it guides itself infallibly to its target. It is not difficult to extrapolate to what the Cruise missile in five to ten years could become – just what it could seek out and differentiate in the terrain below and ahead of it. Coles has discussed the AI aspect of the Cruise missile.[18]

Leaving the military, there are more subtle dangers. The police, or the Mafia, the city administrations, the civil service, the democratic process, the large corporations, the schools, the hospitals, the CIA — all these institutions are already being magnified by computer systems in what they can do, and it is a process which one would expect to snowball. Institutions will take a quantum leap to a new dimension of self-magnification when each organization can invoke the advice of its own growing arsenal of 'expert systems' in the struggle for corporate or political survival.

There is also the possibility that society may be a highly tuned complex control mechanism whose stability absolutely requires built-in delays in its internal transactions. Any control engineer with experience of far less complex mechanisms than the whole of human society knows that such systems are possible. Could it be that human society is one such mechanism? Furthermore the development of telecommunications, of computer science, of computer technology, and ultimately of AI is accelerating. Willed causative actions impinge on their targets and ricochet around in a network which now includes something like 6,000 million individual nodes, i.e. people in the network. Citizens and politicians, and terrorists and police, and administrators can now all react rapidly to the stream of telecommunication information, processing it and reducing it at speed. Are we sure that the entire planetary 'operating system' cannot run itself into all kinds of deadlocks and page-flappings and tear-away processes, responsive to its internal dynamics rather than to the kinds of lives that anyone wants to live?

So much for augmentation of existing things. Now consider novelties, things appearing in the world totally different in principle from any social phenomena that people have ever had to contemplate before. It is here that the rest of the world, the non-AI world which is almost everybody, has its greatest concern about whether scientists can be trusted to be mature about new processes that may be triggered off. The problem is similar in kind to that which arises in the context of genetic manipulation, but I see it as more far-reaching.

The Inscrutable Planet

Let us start on the bad side. A new phenomenon can be predicted by extrapolation from the kind of computer networking which has already added receptor and effector organs to the major control institutions of cities. A city has dozens or scores of highly influential organizations, not just City Hall, but the garbage and sanitary services, the medical organizations, the educational organization, the banks, the airline system, the traffic control system, the building and planning authority and so on. There comes a point when their computing networks begin to talk to each other, initially for quite simple pragmatic reasons. It is helpful when you are reserving your air ticket if the ticketing system can interrogate your bank to see whether your credit is good. It is helpful for the police, looking for somebody known to have had a finger amputation in a certain month in such and such a city, if their computing system can get into the Health Service's computerized hospital records and pull out all the people who had finger amputations at that time.

Obviously there are touchy questions to do with privacy — the rights of the individual to know what information, and to control how much of it, gets swapped around among the files of large organizations. Do they control him or is it the other way round? Has one the right to know what is on one's FBI or KGB or CID file? This article is not concerned with this issue, partly because scientific and political rights are aware of these dangers, and have already shown themselves moderately responsive.

I suggest now that we extrapolate a little: imagine that the city administration network and the medical network and the news media network and the police network and the bank network and the traffic control network are networks of the year 2000 and not of the 1980s, and so have had considerable intelligence incorporated into them, secondly that each network is now communicating quite richly and densely with the other networks. Imagine then an eventual situation in which computer control networks for entire cities have their own goal-setting, in which nobody can be found any more who can understand even the documentation, let alone the systems themselves — only one little pathway in the electronic jungle. We will then have to be certain that all evaluation functions and heuristics are tuned just right, because an electronic city will have to do the normal administrative trading and bargaining with other electronic cities. Each city controls certain resources and can make certain concessions. One city wants something done about the water supply, but another's control system can make that either cheap or expensive for them. In exchange, if the traffic system could be so altered that the football crowds on Saturday go by another route, and so forth . . . Human administrators do this the whole time. Eventually we must envisage bargaining by computer, and bargaining depends critically on perception of values.

One of the values which comes into this bargaining situation, which the Mayor of New York, for example, has to put into the calculation, is the value of human life — how many dollars? — not to mention the value of human values. Such a future scenario might show *homo sapiens* as a parasitic species at best. The year 2500 or the year 3000 could see a planet on which humans are living in the interstices of uncomprehended, incredibly intelligent electronic organisms, like fleas on the backs of dogs. Worse than that, we might become a superseded species, when the dogs ask of the fleas: 'What have you done for us lately?' We have to rely for all future time on the ability of automatic systems to understand the value of human values.

Cultural Immortality

Concerning scenarios in which the species is actually superseded so that we have no genetic descendants, H.A. Simon's view[19] is that this could be borne, because we would have *cultural* descendants — electronic successors to pick up human culture and carry it on. He sees that as a kind of immortality. There are issues here that build a bridge between points of view. There *is* a new kind of immortality. An irreligious person would say that it is the only immortality that humanity has every got near. Consider an aspect of the motivation of the University of Pittsburgh medical project. A distinguished world figure in internal medicine, Dr Myers, has developed diagnostic skills in this area equivalent to those of a Grandmaster in chess. The objective which Harry Pople and he are together pursuing is to transfer, before he dies, his medical skill, his knowledge and inferential strategies, into software.[20] This will be for him a kind of immortality, more vivid and more direct than authorship of medical writings.

The degree to which important facets of a human's intellectual powers and intellectual life can actually be reconstructed and embedded in software will increase. At present it is not fashionable to think that Artificial Intelligence is ever going to make an inroad into less linguistically describable aspects of mental life, the cultural and emotional aspects. Problems which are easy to formalize offer the quickest rewards in AI engineering. But I do not know of any good argument why the more elusive and intuitive skills should not yield to machine implementation in the end. It might be conjectured that there is some hidden barrier which prevents a Bach-understanding project from succeeding, whereas a speech-understanding project *could* succeed. A former colleague of mine, H.C. Longuet-Higgins, is working on computer understanding of music, at present at a necessarily low level — the ability correctly to transcribe keyboard sequences into musical notation. But he has in his long-distance sights a more ambitious goal. To the extent that he does not succeed, there will be people in future generations who may succeed. So the highly tuned discriminatory powers and aesthetic judgements of a music critic may become as easy to build into a knowledge system as diagnostic skills in internal medicine. Fragments of immortality — I would suggest growing fragments of immortality — will be up for sale. We cannot classify such a development as anything but good.

Symbiosis

Assuming that we do *not* want to be superseded, does the idea of man-machine symbiosis offer an alternative? At first sight, brain-computer communication has an aura of fantasy; but at a crude and primitive level experiments have already succeeded.[21] In a project sponsored by ARPA, aircraft pilots were trained to get control over their own electroencephalogram patterns so that by thinking different kinds of thoughts, and thus generating different kinds of EEG patterns, they could output a small alphabet of a dozen or so different signals. The idea was to pick those signals up by computer and process them, interpreting them as pilot's requests for different categories of information, and display the appropriate dial, gauge, etc. on to the VDU screen. The pilot then gets his air-speed indicator when he thinks 'air-speed', altitude when he thinks 'altitude', and so on.

That project reached first-base success. It turned out to be possible. I do not suggest that if brain-computer communication ever becomes big technology it will necessarily go that particular route, monitoring EEG patterns. On the contrary, it seems likely that these ARPA experiments will seem more like the balloonists' ascents of the eighteenth century. These had almost nothing to do with the technology of heavier-than-air flight. But just because they had little to do with it directly, we must not make the mistake of thinking that no contribution was made towards heavier-than-air flight by the balloonists. They had effects in several ways. One was by proving that it could be done without the human aeronaut exploding or having fatal convulsions. People were sceptical about that, just as they were in a later generation about rapid transportation of the human frame by railway. Second, the balloonists created interest among educated people in the idea of flight. Third, they made a start on barometric, navigational and other measurements.

The 'symbiotic' concept points towards our becoming not a superseded species but an augmented one. The experience of the aircraft pilot who can conjure up images from a pseudo-memory must subjectively be very much like conjuring up images from his own memory, getting them flashed up by willing the act of retrieval. Imagine a future with an enormously extended database and processing capacity as an extension of our present intellectual powers, but subjectively built into our mental life in the way that the telescope and the microscope can be, or even the automobile in the case of motor skill. Most people, once they get really accustomed to driving cars, and particularly if they are skilled (for example, professional motor-racing drivers) have a total body sense in which they and the car blend to form a subjective unity. So the extension that the Greeks pictured with their notion of the centaur has now been implemented as a human experience through the technology of automobiles. I would personally extend that image for the ultimate final path that I hope AI will take. I see no reason why it should not.

Motivation and Right of Humans to Survive

Is there any reason for thinking of extension rather than replacement? Is that for sentimental reasons? Is it that human beings simply have a vested interest in maintaining themselves? Any technology that could provide sufficient augmentation could surely in the end dispense with the human part of the compound organism.

It is a question of motivation. If the extended human, the augmented human, is to decide not to reproduce biologically or to find some other way of discarding the human component, there has to be a motive. It would seem to be an unmotivated, or an absurdly motivated, decision.

It is an error to identify our right to exist or our purpose in being here with our mental aptitudes. It is not very common to find athletes thinking: 'The future world's all right . . . so long as there are some athletes around, we can do without all these doctors, and lawyers and mathematicians'. The situation is curiously asymmetrical. Such ideas come more easily to intellectuals: 'We could dispense with these other people. Provided that the intelligent machines of the future keep a few really bright intellectuals around, that's not so bad'. There is a serious error here – not an error in the academic sense, but an error of mortality – in deriving a person's right to make the most of himself from his demonstrated aptitudes.

It may be objected that this is bound to be so because our society is resource-limited. But we must draw a distinction between what is done because it is a pragmatic necessity and what is done because that is the way it should be. There is every reason why we should keep these two things separate. We must be prepared for explicit acceptance that we do all sorts of bad things every day. Every responsible politician or administrator does – and if he is honest he will admit it. He has to do what ideal criteria would classify as bad, because of constraints, time-scales and limited information.

So in all governments there are necessary offices which are not only vile, but vicious too: vices which have there a place, and help to make up the seam in our piecing, as poisons are useful for the preservation of health. If they become excusable, because they are of use to us, and that the common necessity covers their true qualities, we are to resign this part to the most robust and least fearful of the people, who sacrifice their honour and conscience, as others of old sacrificed their lives, for the good of their country, we who are weaker taking upon us the parts that are more easy and less hazardous. The public good requires that men should betray and lie and murder.[22]

But in the idea of progress, which is a very recent one, although things may now have to be done this way we are resolved not to lose sight of a possible world in which they will be ordered differently. A fundamental principle of that possible world should be that a human's right to everything that it takes to develop and express himself is not based on what an assessing tribunal thinks of those self-expressions: whether he is a fine artist or a bad artist, let us say, or whether he is a valuable financial adviser or a distractable drop-out. Either way, he is a human being.

Religions, Myths and Machines

'Will machines have religions?' The answer to that depends on whether God created man, which is one view, or whether the converse, which is that man created God. I personally suppose that it is the second which happened, and that the religions of mankind are part of human culture — that is, they are creations of mankind. Then, the question is: if we have social information-processing systems which generate highly elaborate but instantly recognizable information structures, namely those which we call religious, and if we then notice that they are very similar in all sorts of different ways as between different societies who have independently evolved very similar information-structures of the religious type, then we can expect that the applied mathematics of the future, concerned with the costs and benefits of information-processing, will discover a reason why these myths, mysteries and dogmas are secreted.

There are some obvious conjectures. One is to do with what we are coming in AI to call 'default' reasoning. A religion offers a complete set of slot-fillers. If we have no experimental or deductive grounds for filling certain mental slots, should we criticise the habit of devising convenient conjectures with which to fill them? Will not robots have to have similar default-sets? Perhaps myths are made to plug gaps when one is faced with greater complexity, or prediction difficulty, than one's own explanatory powers or those of one's village or local community can convincingly reach to. Cognition always confronts a dilemma. We feel that explanations and predictions should be rationally grounded. Yet (presumably for good evolutionary reasons) we cannot leave the matter alone. When we lack rational grounds to explain or predict we fill in if necessary with irrational grounds. For example, in an agricultural community: 'When will it rain?' We feel better if we plug vacant slots with made-up explanations. It will rain if the god becomes well disposed. At least we can then try to do something about it (for example, by performing a rain dance). For some temperaments this is the important consideration.

Much depends on the habitual mood of the given community — whether for example the tradition of the community is of an effortful, goal-seeking type. In that case explanations become particularly important. At one end of a spectrum is the American work-oriented paradigm, anxious for achievement, anxious for explanation. At the opposite end lies a society like the Tikopia, a community living in the Pacific in idyllic circumstances.[23] When anthropological study began there was no sign that this little community of 2,000 people had changed in recorded time. Most of their interests centred around gossip, making love, dancing, preparing their midday meal — on which they spent hours — and sometimes a little exploration. Their life was healthy, and by our standards we would say happy. It approximated more closely than one would think possible to Rousseau's idea of the Noble Savage.

But when members of such idyllic communities are asked *why* things happen — things which to us, who weave our mental lives very tightly out of causality, seem to require explanation — they may give fanciful reasons of a kind which I shall caricature as follows: 'Why does the sun go down in the sea every day?' 'It's a big red bird and it wants to go back to its nest.' So then: 'Ah! but if it is a big red bird seeking its nest then why doesn't it stay there? Yet it comes out of the sea the next morning from the other side!' Having thus revealed an entirely wrong mental approach, the questioner may now be requited with something like 'The bird goes where he goes and knows what he knows!'or (ultra-sophisticated) 'Well, maybe the sun isn't a big red bird, then!' These are reasonable attitudes for happy man.

Machines will presumably need myths for the same information-processing reasons that humans need myths. I take a myth to be a belief, which is treated as certain knowledge, and felt desirable that it should be, by some group of interacting information systems. It can sometimes be better to run along with a set of false hypotheses *as though* they were true, if they are quick and computationally cheap. The myth-designer will say: 'Kindly specify what tasks this machine must do, and how much out of its depth, in terms of computational complexity, has it got to operate?' There comes a point when empty belief-slots *must* be plugged with default-values.

Need for a Theory of Lying

Closely related to these considerations is the ultimate necessity of a mathematical theory of lying.

Let us suppose that we have a network of intelligent machines, which have to cooperate with each other. Imagine a group of automation robots in a factory which are to some degree specialized — one does the paint-spraying, another welds, one rivets, and so forth. For each knowledge-base that each robot has, there will be a part which for reasons of economy is common to all the robots. The common part might include facts about the work bench. Not all the robots will want exactly the same facts or the same emphasis. Obviously, if one robot works on top of the bench it may be easier for it if, like the ancients who believed that there was nothing the other side of the world, it does not believe that there is any world beyond the bench top. That is a kind of myth, and there is no point whatsoever in the robot having in its knowledge-base the information that this *is* a myth. The supervisory robot might as well let it believe the myth and save core-store. The attempt to imagine control systems for cooperating intelligent devices, even at the mundane level of assembly-line or factory-floor situations, is brought up against something which we knew already but have not hitherto thought important. I refer to the fact that if I want to influence inanimate matter in a highly predictable way then I have physics. But suppose that the object that I want to influence is itself an information-processing system, a human being, then I may *either* seek to influence the system's actions like a military general by issuing impera-tives, *or* I may make assertional statements, having a good enough model of the target system to know that these will have the same effect as the imperatives, but in some circumstances will act more quickly and cheaply.

Consider as an example the economy practised by the British road-sign authorities. A typical road sign has a picture on it somewhat like that in Figure 29.4. Everybody instantly interprets it as saying: 'When you get onto the roundabout, go clockwise'. But that is not actually what the picture says. The picture presents its viewer with an *assertion*, namely that there is a defect, a gap in the road. And this is a lie. Strictly speaking it is a joke lie, since there is no intent to deceive. But those who invented the sign have a good enough model of a motorist to know — first that if he believes it, he will go round to the left; second that if he does not believe it, and knows it to be false, he will be subtle enough to perform an addi-tional inference and realize that the *intention* is to give him an imperative.

The same phenomenon is exemplified when parents tell children absurd untruths, as politicians also do voters, on the hypothesis that this is the easiest way to get them to do something, or not to do something: 'It will make you sick!', meaning 'Don't eat it!' in the private knowledge that it will *not* make the child sick. Like the motorist, the child usually knows this too. But because time is too short either to have a clash of wills or alternatively to explain the real reasons, this short-cut is tried — and it is in assertional language.

Hence, we, or the theorists of knowledge-engineering among our descendants, will eventually be obliged to devise an economics of lying. Presumably lying has arisen following regular economic laws of costs and benefits. Presumably we shall be bound by the same laws to construct analogues in intelligent machines.

FIGURE 27.4.   Scheme used in road signs to indicate desired direction round a roundabout. Strictly, the depicted defect in the road is a falsehood, but the motorist understands what is required of him. (Photography courtesy of J.K. Wilkie, Edinburgh.)

Promotion and Control of Research

Enough has been said to establish that it is in the human interest, conceived narrowly in terms of industrial technology, science, medicine, education and the propagation of culture generally, that the new development should be fostered. Enough has also been said to indicate that we are sitting on a powder keg. Hence a broad interpretation of social responsibility indicates that events should to some degree be monitored and the development regularized.

A minimal first step should be a properly equipped, funded and managed facility available to those scientists who wish to pursue research, or just to gain familiarity and competence, in the new methodologies. No national laboratory for long-range computing research exists in Britain. The lack not only retards AI. Cognate branches of experimental computation suffer, for example those related to automatic programming. In these ways damage is compounded, and our nation's apparently self-chosen role of technological and cultural dependency moves closer. An indirect comment on past indecision, and wrong decision, can be found in the early history of aeronautics.

On 28 January 1909 Lord Esher's sub-committee on the Committee of Imperial Defence recommended the withdrawal of support and facilities from the pioneering work on heavier-than-air flight then building up at Farnborough. At the parent committee's following meeting the Prime Minister, H.H. Asquith, 'expressed his satisfaction that work on aeroplanes was to cease, and ruled that the sub-committees's recommendations be accepted'. The following is the continuation of this quotation from Mr P.B. Walker's authoritative history:[24]

This should have been the end, but, as is well known, the British public were not happy with such a ruling, especially after Louis Bleriot's sensational arrival by air from France on 25 July 1909. Concern as regards the present History, however, is with the haunting memory of the sub-committee's findings. Lord Esher was not only a man of high intelligence and influence, but also a man with a conscience and moral courage. He realised that his committee had made an appalling mistake, and in October 1910 he wrote to the Committee of Imperial Defence telling them so. In a note dealing in some detail with Britain's aviation problems generally, he expressed the view that Britain must arm herself with a whole fleet of aeroplanes. Unless she did so, he said, she would be in mortal peril, and for this the Imperial Defence Committe would be rightly blamed.

It should be noted however that no obstacles had been placed by the Imperial Defence Committee in the way of individuals in Britain's universities had they desired to pursue aero-engineering with the normal supports available to academic laboratories. In the case of aeronautics this circumstance was plainly irrelevant. Since in the new context of computing the matter has to be argued afresh, the case will be briefly summarized here.

Until the early 1970s university-based research in the United Kingdom was supported by the Department of Education and Science according to a pattern known as the 'dual support system'. The theory, which corresponded fairly closely with actual practice, went as follows.

A university is an institution formed to promote two mutually beneficial objectives, namely teaching and research. Consequently it devotes its resources of salaries, building, equipment and services to the enablement of both these activities. It is recognized, however, that modern science is developing on a scale which makes it not always possible for a university to meet the full requirements of every timely and promising research programme. To meet the short-fall the Research Councils invite applications from tenured staff in universities for scientific investigations which need to be 'topped up' with additional resources. Proposals judged by peer review to be meritorious thus become eligible for grants in aid. Research Council grants defray such costs of staff and equipment as are beyond a university's power to supply. But the university *is* asked to bear the entire added burden imposed by the 'overheads' of each new project — typically amounting to half the value of the grant again — financial and clerical administration, accommodation, heating, lighting, telephones, mail, reprographics and other documentation costs and consumables. When the grant runs to its close, the university is also expected to assimilate on to its own permanent salaried staff most of those scientists who were hired on the grant.

This system was workable, and on occasion worked impressively, in the buoyant conditions of the 1960s, the decade of the Robbins university expansion and the 'white-hot technological revolution'. In the early 1970s the dual support system collapsed, its fundamental assumptions shattered beyond repair by the economic crisis.

The reactions of the Research Councils varied. The Medical Research Council, in recognition of the demise of 'dual support', augmented its existing arrangements. In particular the MRC now itself covers in large measure the 'overhead' costs of its grants. In addition, in selected cases it has introduced new categories of especially long-lasting support — in some cases for periods as great as ten years.

Computing research, on the other hand, comes under the aegis of the Science Research Council. This body has taken a different view of the dual support system — namely that until such time as it is officially pronounced dead it must continue to be the basis. If present circumstances make it impossible for the system to function then there is no alternative but to carry on *as though* it were operative. This in itself, it is felt, may bring pressure to bear on universities to invest something at least in the underpinnings of research.

In the outcome, small-scale and short-term studies in computing science flourish as before. But long-range investigations of the kind which require stability and major institutional commitments have rather abruptly vanished from the scene. In the framework just described there is no way of restarting them.

The relevant branches of computer science are those which are heavily dependent on large-scale experiment and incremental construction of advanced instrumentation, including complex software. Two interrelated research themes of this kind are (1) expert systems and (2) 'learning' robots and self-programming computers. The case for establishing at least one national institute is that such work can then be re-started.

Conclusions

Artificial Intelligence, one of the most recently developed branches of science, is becoming an economic reality. 'Expert systems' can provide knowledgeable advice on demand in selected areas of consultancy. Acceleration of the new technology can be expected to follow laboratory solutions to problems of machine learning, while additional potential has been discovered in the use of such systems to improve the quality of traditional knowledge sources. Central to all applications is the need for 'humanized' machine representations which allow mutually intelligible interaction. Such representations must encompass probabilistic as well as logical systems of inference. A major past forecast of rates of progress in AI has so far been corroborated by subsequent events. It envisaged attainment of human-level all-round intelligence in machines by the early twenty-first century.

Social benefits could include insightful information retrieval for scientists, scholars and technologists; and for the consumer a many-faceted up-grading over almost the entire range of material goods. Aside from military applications, dangers can be seen in loss of stability in social control mechanisms, increasing dependency and decreasing comprehension on the part of the user population and loss of independent will and vital impetus by mankind at large. A possible line of solution, leading to greatly augmented amplitude of life, may conceivably be found through direct man-machine symbiosis. It will certainly be unwise for our species to apply the historical pattern of encouraging the extinction of less viable groups. In a future in which the planet has to be shared with intelligent machines, perpetuation of the principle could become two-edged.

Technological pressure is likely to lead to incorporation into intelligent devices of certain mental patterns regarded by strict rationalists as mere foibles. Included in this category are religions, myths and the practice of lying

So that further study of AI and other socially critical areas of computing can be more effectively promoted and regularized, it would be desirable to establish a central facility for research in advanced computation.

REFERENCES

1. Hart, P.E., Duda, R.O. and Einaudi, M.T. (1978). 'PROSPECTOR – A computer-based consultation system for mineral exploration', *Math. Geology*, Vol. 10, pp. 589–610.
2. De Dombal, F.T., Leaper, D.J., Staniland, J.R. *et al*. (1972). 'Computer-aided diagnosis of acute abdominal pain', *Brit. Med. J.*, Vol. 2, pp. 9–13. Walmsley, G.L., Wilson, D.H., Gunn, A.A., Jenkins, D., Horrocks, J.C. and De Dombal, F.T. (1977). 'Computer-aided diagnosis of lower abdominal pain in women', *Brit. J. Surg.*, Vol. 64, pp. 538–41.
3. Michie, D. and Chambers, R.A. (1968). 'BOXES: an experiment in adaptive control', *Machine Intelligence 2* (eds. Dale, E. and Michie, D.), Edinburgh: Edinburgh University Press, pp. 137–52.
4. Ambler, A.P., Barrow, H.G., Brown, C.M., Burstall, R.M. and Popplestone, R.J. (1975). 'A versatile system for computer-controlled assembly', *Artificial Intelligence*, Vol. 6, pp. 129–56.
5. Shortliffe, E.H. (1976). *Computer-Based Medical Consultations: MYCIN*, New York: Elsevier.
6. Yu, V.L., Fagan, L.M., Wraith, S.M. *et al*. (1979). 'Antimicrobial selection for meningitis by a computerized consultant – a blinded evaluation by infectious disease experts', *Journal of the American Medical Association*, Vol. 242, pp. 1279–82.
7. Bennet, J., Creary, L., Engelmore, R.S. and Mebosh, R. (1978). 'SACON: a knowledge-based consultant for structural analysis', *Memo HPP-78-28*, also *Report No. STAN-CS-78-699* Stanford: Computer Science Department, Stanford University.
8. Nievergelt, J. (1977). 'The information content of a chess position, and its implications for the chess-specific knowledge of chess players', *SIGART Newsletter*, Vol. 62, pp. 13–15.
9. Hayes, J.E. (1978). 'Children's visual descriptions', *Cognitive Science*, Vol. 2, pp. 1–15.
10. Michie, D. (1977). 'A theory of advice', *Machine Intelligence*, Vol. 8 (eds. Elcock, E.W. and Michie, D.), Chichester: Ellis Horwood and New York: Wiley (Halstead Press), pp. 151–68.
11. Good, I.J. (1950). *Probability and the Weighing of Evidence*, London: Griffin.
12. Michie, D. (1976). 'Bayes, Turing and the logic of corroboration', *AISB European Newsletter*, Vol. 23, pp. 33–6.
13. Firschein, O., Fischler, M.A., Coles, L.S. and Tenenbaum, J.M. (1973). 'Forecasting and assessing the impact of artificial intelligence on society', *Third Int. Joint Conf. on Art. Intell.*, pp. 105–20, Menlo Park: SRI Publications Department.

14. *Social Survey Report No. 245* (1959). 'The use of technical literature by industrial technologists', London: Central Office of Information. Reviewed by C. Scott in *Discovery*, Vol. 20, March 1959, pp. 110–14.
15. Wells, H.G. (1938). *World Brain*, London: Methuen.
16. Newell, A. 'Fairytales', *Viewpoints*, No. 3, Pittsburgh: Carnegie-Mellon University, 1976. Also in *Firbush News*, Vol. 8, pp. 25–9, Edinburgh: Machine Intelligence Research Unit, 1978.
17. Tsipis, K. (1977). 'Cruise missiles', *Scient. Amer.*, Vol. 236, No. 2, pp. 20–9.
18. Coles, L.S. (1977). 'Military funding of AI research', *AISB European Newsletter*, No. 25, pp. 10–14.
19. Simon, H.A. Carnegie-Mellon University, personal communication.
20. Pople, H.E., Myers, J.D. and Miller, R.A. (1977). 'DIALOG: a model of diagnostic logic in internal medicine', *Fifth Int. Joint Conf. Art. Intell.* Pittsburgh: Computer Science Dept., Carnegie-Mellon University.
21. Crocker, S., now at University of California at Los Angeles, USA, personal communication.
22. Montaigne, Michel de (1958). Essay No. 1, Third Book (Hazlitt's translation).
23. Firth, R. (1936). *We the Tikopia: a Sociological Study of Kinship in Primitive Polynesia*, London: George Allen & Unwin. Republished 1963, Beacon Paperback (USA).
24. Walker, P.B. (1974). *Early Aviation at Farnborough: The History of the Royal Aircraft Establishment. Volume II, The First Aeroplanes,* London: Macdonald, pp. xxi–xxii.

Subject Index

307

Author Index